# HITE 7.0
# 培养体系

HITE 7.0全称厚溥信息技术工程师培养体系第7版，是武汉厚溥企业集团推出的“厚溥信息技术工程师培养体系”，其宗旨是培养适合企业需求的IT工程师，该体系被国家工业和信息化部人才交流中心鉴定为国家级计算机人才评定体系，凡通过HITE课程学习成绩合格的学生将获得国家工业和信息化部颁发的“全国计算机专业人才证书”，该体系教材由清华大学出版社全面出版。

HITE 7.0是厚溥最新的职业教育课程体系，该职业体系旨在培养移动互联网开发工程师、智能应用开发工程师、企业信息化应用工程师、网络营销技术工程师等。它的独特之处在于每年都要根据技术的发展进行课程的更新。在确定HITE课程体系之前，厚溥技术中心专业研究员在IT领域和一些非IT公司中进行了广泛的行业调查，以了解他们在



的数据库系统、前端

序，每个产品系列均

工程师为目标而设计

IT行业的岗位序列做了充分的调研，包括研究从业人员技术方向、项目经验和职业素质等方面的需求，通过对所面向学生的自身特点、行业需求的现状以及项目实施等方面的详细分析，结合厚溥对软件人才培养模式的认知，按照软件专业总体定位要求，进行软件专业产品课程体系设计。该体系集应用软件知识和多领域的实践项目于一体，着重培养学生的熟练度、规范性、集成和项目能力，从而达到预定的培养目标。整个体系基于ECDIO工程教育课程体系开发技术，可以全面提升学生的价值和学习体验。

## 一、移动互联网开发工程师

在移动终端市场竞争下，为赢得更多用户的青睐，许多移动互联网企业将目光瞄准在应用程序创新上。如何开发出用户喜欢，并能带来巨大利润的应用软件，成为企业思考的问题，然而这一切都需要移动互联网开发工程师来实现。移动互联网开发工程师成为求职市场的宠儿，不仅薪资待遇高，福利好，更有着广阔的发展前景，倍受企业重视。

移动互联网企业对Android和Java开发工程师需求如下：

| 已选条件： | Java(职位名) | Android(职位名) |
|---|---|---|
| 共计职位： | 共51014条职位 | 共18469条职位 |

### 1. 职业规划发展路线

| Android | | | | |
|---|---|---|---|---|
| ★ | ★★ | ★★★ | ★★★★ | ★★★★★ |
| 初级Android<br>开发工程师 | Android<br>开发工程师 | 高级Android<br>开发工程师 | Android<br>开发经理 | 移动开发<br>技术总监 |
| **Java** | | | | |
| ★ | ★★ | ★★★ | ★★★★ | ★★★★★ |
| 初级Java<br>开发工程师 | Java<br>开发工程师 | 高级Java<br>开发工程师 | Java<br>开发经理 | 技术总监 |

## 2. 素质能力提升路径

| 1 大学生 | 2 大学生活 | 3 学习习惯 | 4 职业目标 | 5 沟通表达 | 6 自我管理 |
|---|---|---|---|---|---|
| 12 准职业人 | 11 职业路线 | 10 求职技能 | 9 就业意识 | 8 融入团队 | 7 形象礼仪 |

## 3. 专业技能提升路径

| 1 大学生 | 2 计算机基础 | 3 编程基础 | 4 软件工程 | 5 数据库 | 6 网站技术 |
|---|---|---|---|---|---|
| 12 准职业人 | 11 产品规划 | 10 项目技能 | 9 高级应用 | 8 APP开发 | 7 基础应用 |

## 4. 项目介绍

(1) 酒店点餐助手

(2) 音乐播放器

# 二、智能应用开发工程师

随着物联网技术的高速发展，我们生活的整个社会智能化程度将越来越高。在不久的将来，物联网技术必将引起我国社会信息的重大变革，与社会相关的各类应用将显著提升整个社会的信息化和智能化水平，进一步增强服务社会的能力，从而不断提升我国的综合竞争力。 智能应用开发工程师未来将成为热门岗位。

智能应用企业每天对.NET开发工程师需求约15957个岗位(数据来自51job)：

| 已选条件： | .NET(职位名) |
|---|---|
| 共计职位： | 共15957条职位 |

## 1. 职业规划发展路线

| ★ | ★★ | ★★★ | ★★★★ | ★★★★★ |
|---|---|---|---|---|
| 初级.NET开发工程师 | .NET开发工程师 | 高级.NET开发工程师 | .NET开发经理 | 技术总监 |
| ★ | ★★ | ★★★ | ★★★★ | ★★★★★ |
| 初级开发工程师 | 智能应用开发工程师 | 高级开发工程师 | 开发经理 | 技术总监 |

## 2. 素质能力提升路径

| 1 大学生 | 2 大学生活 | 3 学习习惯 | 4 职业目标 | 5 沟通表达 | 6 自我管理 |
|---|---|---|---|---|---|
| 12 准职业人 | 11 职业路线 | 10 求职技能 | 9 就业意识 | 8 融入团队 | 7 形象礼仪 |

## 3. 专业技能提升路径

| 1 大学生 | 2 计算机基础 | 3 编程基础 | 4 软件工程 | 5 数据库 | 6 网站技术 |
|---|---|---|---|---|---|
| 12 准职业人 | 11 产品规划 | 10 项目技能 | 9 高级应用 | 8 智能开发 | 7 基础应用 |

## 4. 项目介绍

(1) 酒店管理系统

(2) 学生在线学习系统

# 三、企业信息化应用工程师

当前，世界各国信息化快速发展，信息技术的应用促进了全球资源的优化配置和发展模式创新，互联网对政治、经济、社会和文化的影响更加深刻，围绕信息获取、利用和控制的国际竞争日趋激烈。企业信息化是经济信息化的重要组成部分。

IT企业每天对企业信息化应用工程师需求约11248个岗位（数据来自51job）：

| 已选条件: | ERP实施(职位名) |
|---|---|
| 共计职位: | 共11248条职位 |

## 1. 职业规划发展路线

| 初级实施工程师 | 实施工程师 | 高级实施工程师 | 实施总监 |
|---|---|---|---|
| 信息化专员 | 信息化主管 | 信息化经理 | 信息化总监 |

## 2. 素质能力提升路径

| 1 大学生 | 2 大学生活 | 3 学习习惯 | 4 职业目标 | 5 沟通表达 | 6 自我管理 |
|---|---|---|---|---|---|
| 12 准职业人 | 11 职业路线 | 10 求职技能 | 9 就业意识 | 8 融入团队 | 7 形象礼仪 |

## 3. 专业技能提升路径

| 1 大学生 | 2 计算机基础 | 3 编程基础 | 4 软件工程 | 5 数据库 | 6 网站技术 |
|---|---|---|---|---|---|
| 12 准职业人 | 11 产品规划 | 10 项目技能 | 9 高级应用 | 8 实施技能 | 7 基础应用 |

## 4. 项目介绍

(1) 金蝶K3

(2) 用友U8

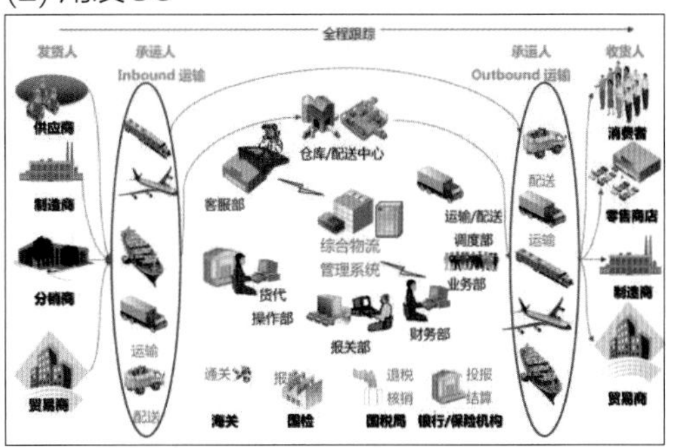

## 四、网络营销技术工程师

在信息网络时代，网络技术的发展和应用改变了信息的分配和接收方式，改变了人们生活、工作、学习、合作和交流的环境，企业也必须积极利用新技术变革企业经营理念、经营组织、经营方式和经营方法，搭上技术发展的快车，促进企业飞速发展。网络营销是适应网络技术发展与信息网络时代社会变革的新生事物，必将成为跨世纪的营销策略。

互联网企业每天对网络营销工程师需求约47956个岗位(数据来自51job)：

| 已选条件: | 网络推广SEO(职位名) |
| --- | --- |
| 共计职位: | 共47956条职位 |

### 1. 职业规划发展路线

| 网络推广专员 | 网络推广主管 | 网络推广经理 | 网络推广总监 |
| --- | --- | --- | --- |
| 网络运营专员 | 网络运营主管 | 网络运营经理 | 网络运营总监 |

### 2. 素质能力提升路径

| 1 大学生 | 2 大学生活 | 3 学习习惯 | 4 职业目标 | 5 沟通表达 | 6 自我管理 |
| --- | --- | --- | --- | --- | --- |
| 12 准职业人 | 11 职业路线 | 10 求职技能 | 9 就业意识 | 8 融入团队 | 7 形象礼仪 |

### 3. 专业技能提升路径

| 1 大学生 | 2 计算机基础 | 3 编程基础 | 4 网站建设 | 5 数据库 | 6 网站技术 |
| --- | --- | --- | --- | --- | --- |
| 12 准职业人 | 11 产品规划 | 10 项目实战 | 9 电商运营 | 8 网络推广 | 7 网站SEO |

### 4. 项目介绍

(1) 品牌手表营销网站

(2) 影院销售网站

HITE 7.0 软件开发与应用工程师

# SSM 企业级开发框架

武汉厚溥数字科技有限公司　编著

清華大学出版社

北　京

## 内 容 简 介

本书按照高等院校计算机课程的基本要求进行设计，注重理论和实践结合，采用先实践再总结的方式，突出计算机课程的实践性特点。本书共包括 16 个单元：SSM 框架简介，MyBatis 核心组件，MyBatis 核心配置，映射器，动态 SQL，Bean 工厂与应用上下文，Spring AOP 编程，Spring 和数据库编程，Spring 数据库事务管理，Spring MVC 的开发流程，Spring MVC 组件开发，Spring MVC 项目开发，业务的抽象，前后端交互，页面美化，复用与结尾等。

本书内容安排合理，结构清晰，通俗易懂，实例丰富，可作为各类高等院校、培训机构的教材，也可供大数据程序开发人员参考。



图书在版编目(CIP)数据

SSM 企业级开发框架 / 武汉厚溥数字科技有限公司编著. —北京：清华大学出版社，2023.2
(HITE 7.0 软件开发与应用工程师)
ISBN 978-7-302-61983-3

I. ①S… II. ①武… III. ①企业—计算机网络 IV. ①TP393.18

中国版本图书馆 CIP 数据核字(2022)第 175622 号

**责任编辑：**刘金喜
**封面设计：**王 晨
**版式设计：**孔祥峰
**责任校对：**成凤进
**责任印制：**曹婉颖

**出版发行：**清华大学出版社
**网 址：**http://www.tup.com.cn，http://www.wqbook.com
**地 址：**北京清华大学学研大厦 A 座 **邮 编：**100084
**社 总 机：**010-83470000 **邮 购：**010-62786544
**投稿与读者服务：**010-62776969，c-service@tup.tsinghua.edu.cn
**质 量 反 馈：**010-62772015，zhiliang@tup.tsinghua.edu.cn
**印 装 者：**三河市铭诚印务有限公司
**经 销：**全国新华书店
**开 本：**185mm×260mm **印 张：**21 **彩 插：**2 **字 数：**433 千字
**版 次：**2023 年 2 月第 1 版 **印 次：**2023 年 2 月第 1 次印刷
**定 价：**89.00 元

---

产品编号：099256-01

# 编 委 会

# 前言

Java SSM 框架即指 Spring、Spring MVC、MyBatis，框架集由 Spring、MyBatis 两个开源框架整合而成(Spring MVC 是 Spring 中的部分内容)，常作为数据源较简单的 Web 项目的框架。其中 Spring 框架是一系列当下主流应用框架的核心，是整合其他应用框架的基础，具有非入侵式的设计、轻量级、方便解耦、易于上手等诸多优点。Spring MVC 是 Spring 框架的一个重要模块，是基于 MVC 开发模式的轻量级的 Web 框架。MyBatis 是一款基于 Java 语言的持久层框架，是一款半自动的 ORM 映射工具，可以自定义高效的 SQL 语句，有效提升对数据库的操作效率。

本书是“工信部国家级计算机人才评定体系”中的一本专业教材。“工信部国家级计算机人才评定体系”是由武汉厚溥数字科技有限公司开发，以培养符合企业需求的软件工程师为目标的 IT 职业教育体系。在开发该体系之前，我们对 IT 行业的岗位序列做了充分的调研，包括研究从业人员技术方向、项目经验和职业素养等方面的需求，通过对所面向学生的特点、行业需求的现状以及项目实施等方面的详细分析，结合我公司对软件人才培养模式的认知，按照软件专业总体定位要求，进行软件专业产品课程体系设计。该体系集应用软件知识和多领域的实践项目于一体，着重培养学生的熟练度、规范性、集成和项目开发能力，从而达到预定的培养目标。

本书共包括 16 个单元：SSM 框架简介，MyBatis 核心组件，MyBatis 核心配置，映射器，动态 SQL，Bean 工厂与应用上下文，Spring AOP 编程，Spring 和数据库编程，Spring 数据库事务管理，Spring MVC 的开发流程，Spring MVC 组件开发，Spring MVC 项目开发，业务的抽象，前后端交互，页面美化，复用与结尾等。

本书的编写体系经过编者们的精心设计，按照“理论学习—知识总结—上机操作—课后习题”这一思路进行编排。“理论学习”部分描述通过案例要达到的学习目标与涉及的相关知识点，使学习目标更加明确；“知识总结”部分概括案例所涉及的知识点，使知识点得以完整系统地呈现；“上机操作”部分对案例进行了详尽分析，通过完整的步骤帮助读者快速掌握该案例的操作方法；“课后习题”部分帮助读者理解章节的知识点。本书在内容编写方面，力求细致全面；在文字叙述方面，注意言简意赅、重点突出；在案例选取方面，强调案例的针对性和实用性。

本书凝聚了编者多年来的教学经验和成果，可作为各类高等院校、培训机构的教材，也可供广大程序设计人员参考。

本书由武汉厚溥数字科技有限公司编著，由翁高飞、寇立红、王伟、陈惺、宋晓光、李伟、冯玲、陈红志、邓洁、方海根等多名企业实战项目经理编写。本书编者长期从事项目开发和教学实施，并且对当前高校的教学情况非常熟悉，在编写过程中充分考虑到不同学生的特点和需求，加强了项目实战方面的内容。本书编写过程中，得到了武汉厚溥数字科技有限公司各级领导的大力支持，在此对他们表示衷心的感谢。

参与本书编写的人员还有：河南财政金融学院的席红旗，张家口机械工业学校的冯景蕊、柯玉军，江西青年职业学院的朱烜伯，陕西机电职业技术学院的张克，闽西职业技术学院的黄素叶，内蒙古能源职业学院的张瑞英等。

限于编写时间和编者的水平，书中难免存在不足之处，希望广大读者批评指正。

服务邮箱：476371891@qq.com。

编　者

2022 年 9 月

# 目　录

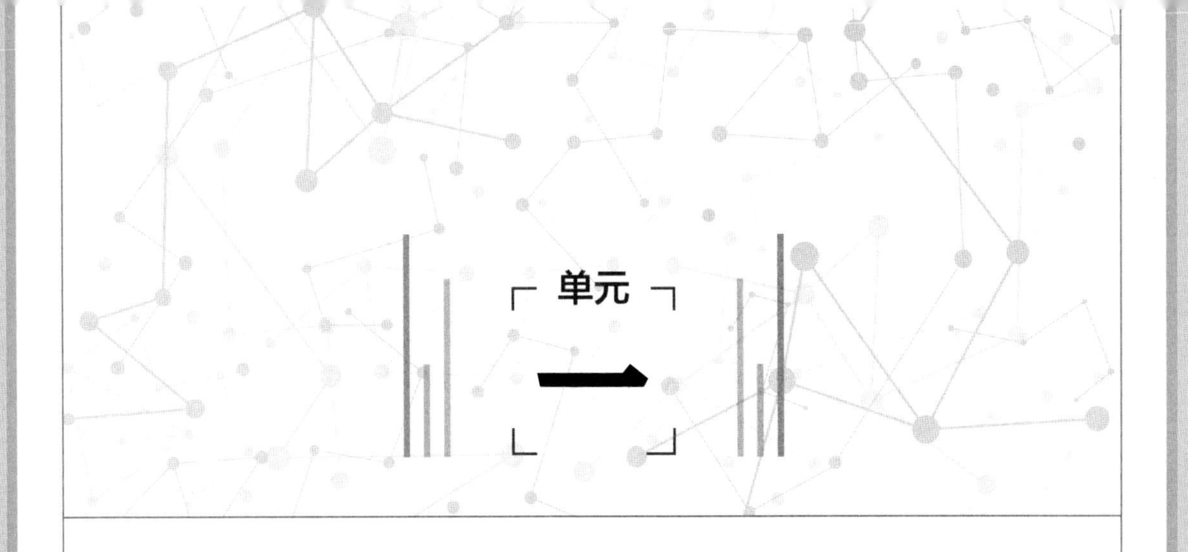


# SSM框架简介

## 课程目标

- ❖ 了解 Java 企业级应用及 SSM 框架
- ❖ 理解 Spring 轻量级容器
- ❖ 了解 Spring 5 的新特性

**简介**

SSM(Spring+Spring MVC+MyBatis)框架是Spring和MyBatis两个开源框架的整合(Spring MVC是Spring中的部分内容)，是标准的MVC模式，它将整个系统划分为View层、Controller层、Service层、DAO层四层。使用Spring MVC负责请求的转发和视图管理，Spring实现业务对象管理，MyBatis作为数据对象的持久化引擎。

# 1.1 Spring框架

Spring是一个开源框架，Spring是于2003 年兴起的一个轻量级的Java 开发框架，由Rod Johnson 在其著作Expert One-On-One J2EE Development and Design中阐述的部分理念和原型衍生而来。它是为了解决企业应用开发的复杂性而创建的。

## 1.1.1 Spring轻量级容器

在目前的企业级开发中，应用程序都需要一个基础骨架，也需要管理业务对象，而提到业务对象的管理，通常不能不提容器这个概念。所谓容器，是指程序代码的运行框架，应用程序对象(多是业务对象)在容器中运行，这也就是通常所说的“被容器管理”。有很多以容器为基础的架构和模型，比如EJB的架构。EJB是用于开发和部署多层结构的、分布式的、面向对象的Java应用系统的跨平台的构件体系结构，在这种架构中，EJB就是一个容器，它用于管理J2EE业务对象。

通常情况下，容器应该提供以下常见的功能。

(1) 生命周期：容器用于控制应用程序对象的生存周期。

(2) 查找服务：容器提供方法用于获得被容器管理的业务对象的引用，这是容器的核心功能。

(3) 配置管理：提供统一的方法使得在不修改源代码的情况下来配置运行在容器中的对象。

(4) 依赖决议：除了简单类型的配置外，容器还可以管理各个业务对象之间的关系。

(5) 企业级服务：为容器内运行的对象提供声明性事务或声明性安全等J2EE开发中常见的服务。

(6) 线程管理：为运行在容器中的对象提供线程模型。

使用容器可以简化 J2EE 应用程序的开发，原因是：

(1) 可接插性，解决业务对象间的依赖问题：容器可以为各种组件提供接插性，将调用方与具体的实现隔离开来，实现按接口进行编程，并且提供了某种方法来找到接口的具体实现，避免在代码中进行硬编码。

(2) 服务定位的一致性：容器提供方法用于获得被容器管理的业务对象的引用。

(3) 配置管理功能：容器提供统一的方法使得在不修改源代码的情况下可以配置运行在容器中的对象。

(4) 提供诸如事务、安全等企业级服务。

容器非常重要，可以提高生产率，让开发者分离关注点，将更多精力放在业务梳理和实现上。当然，应用程序也可以不使用 EJB 的特定架构，不使用容器来管理业务对象。由于没有使用容器，所以无法使用容器提供的种种优点，例如为了实现业务对象的配置，需要编写大量的工厂类，用来处理各种各样的配置，这样会导致项目整体的代码有很大的差异，带来更高的维护成本。再如，没有使用容器，那些本来由容器提供的企业级服务也需要由程序员写代码来实现。

EJB 在项目中量级比较重，本身也比较复杂，学习成本较高，而且出现问题也不容易解决，有很多潜在问题。所幸的是，我们还有另一种方案，可以在使用容器所有优点的同时，避免使用重量级容器(比如 EJB)带来的种种弊端，这就是现在比较流行的框架技术——轻量级容器。轻量级容器也是一种容器，它可以提供容器的所有功能，容器中所有的对象都是 POJO(Plain Ordinary Java Object)。POJO 是简单的 Java 对象，实际就是普通的 JavaBeans，是为了避免和 EJB 混淆所创造的简称。相比 EJB 容器来说，它又有诸多的优点：

(1) 代码可以脱离容器。与 EJB 容器中的对象必须使用特定的接口或 API 不同，POJO 本身的无侵入性使得代码可以脱离容器独立存在。

(2) 相对于 EJB 容器来说，轻量级容器部署方便，启动迅速。

(3) EJB 容器只能管理粗粒度对象，轻量级容器则可以管理细粒度对象，可以更好地面向对象。

(4) 在业务对象依赖方面，EJB 容器本身并不管理 EJB 组件之间的依赖关系，仅提供 JNDI 方式进行定位，必须自己写代码通过 JNDI 来获得其他 EJB 组件。而轻量级容器大多实现了依赖注入模式，可以很轻松地通过配置文件来解决对象间的依赖问题，避免了 EJB 中使用的硬编码的方式。

(5) 提高了可测试性：由于所有的对象都是 POJO，不依赖于特定的接口和 API，所以可以在不启动容器的前提下进行测试，同时通过依赖注入，使得编写测试的桩和驱动模块更加轻松。

### 1.1.2 Spring 简介

当前有很多免费和开源的轻量级容器，最流行的也是使用得最多的是 Spring。并且围绕Spring 核心衍生了一系列框架，如 Spring Boot、Spring Cloud、Spring Data、Spring Security、Spring Web Flow、Spring Batch 等。这些优秀的框架带给开发者越来越多的便利。本节介绍的是 Spring 框架本身。

Spring 是一个流行的轻量级的 J2EE 开源应用程序框架，其核心代码均来自于真实项目，它的目的是要解决企业级应用程序开发的复杂性，虽然它从 2003 年才出现，但目前已经对 Java 企业级应用体系产生了前所未有的冲击，其核心理念是：

(1) 简化 J2EE 的开发。Spring 是一套基础框架，它能很方便地帮助开发人员更加简单和合理地使用 J2EE 服务，使开发人员能够集中精力实现业务逻辑和其他应用功能，从而降低开发难度，提高开发效率，使代码维护更加容易，且降低开发成本。

(2) 整合各类框架，易于选择。Spring 提供“一站式”的框架整合方案，涵盖了 J2EE 软件开发中的各个分层，包括持久层、业务层、表示层等。在各个分层解决方案中，它本身并未重建新的解决方案，而是和自己的子框架或者现有的开源框架进行无缝结合，这些子框架或开源项目之间彼此可以独立，也可以使用其他框架方案加以替代，使得企业级软件开发人员可以进行灵活的选择。

(3) 统一对象依赖的配置方式。在 Spring 中建立对象之间的依赖关系，不需要自制的对象工厂，都可使用配置文件的方式来完成，提供了一个统一的配置方式。

(4) 非侵入性，使得测试更加方便。

组成 Spring 框架的每个模块(或组件)都可以单独存在，或者与其他一个或多个模块联合实现，如图 1-1 所示。

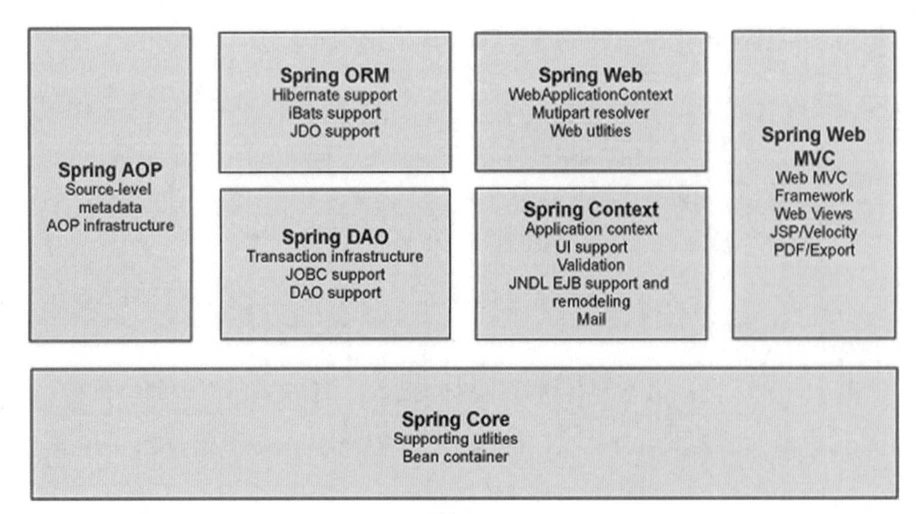


图 1-1

每个模块的功能如下。

(1) 核心容器(Spring Core)：核心容器提供 Spring 框架的基本功能。核心容器的主要组件是 BeanFactory，它是工厂模式的实现。BeanFactory 使用控制反转(IoC)模式将应用程序的配置和依赖性规范与实际的应用程序代码分开。

(2) Spring 上下文(Spring Context)：Spring 上下文是一个配置文件，向 Spring 框架提供上下文信息。Spring 上下文包括诸如 JNDI、EJB、电子邮件、国际化、校验和调度功能等企业服务。

(3) Spring AOP：通过配置管理特性，Spring AOP 模块直接将面向切面的编程功能集成到了 Spring 框架中。所以，可以很容易地使 Spring 框架管理的任何对象支持 AOP。Spring AOP 模块为基于 Spring 的应用程序中的对象提供了事务管理服务。通过使用 Spring AOP，不用依赖 EJB 组件，就可以将声明性事务管理集成到应用程序中。

(4) Spring DAO：JDBC DAO 抽象层提供了有意义的异常层次结构，可用该结构来管理异常处理和不同数据库供应商抛出的错误消息。异常层次结构简化了错误处理，并且极大地降低了需要编写的异常代码数量(例如打开和关闭连接)。Spring DAO 的面向 JDBC 的异常遵从通用的 DAO 异常层次结构。

(5) Spring ORM：Spring 框架插入了若干个 ORM(Object Relational Mapping，对象一关系映射)框架，从而提供了 ORM 的对象关系工具，其中包括 JDO、Hibernate 和 iBatis SQL Map。所有这些都遵从 Spring 的通用事务和 DAO 异常层次结构。

(6) Spring Web 模块：Web 上下文模块建立在应用程序上下文模块之上，为基于 Web 的应用程序提供了上下文。所以，Spring 框架支持与 Jakarta Struts 的集成。Web 模块还简化了处理多部分请求以及将请求参数绑定到域对象的工作。

(7) Spring MVC 框架：MVC 框架是一个全功能的构建 Web 应用程序的 MVC 实现。通过策略接口，MVC 成为高度可配置的框架，MVC 容纳了大量视图技术，其中包括 JSP、Velocity、Tiles、iText 和 POI。

Spring 框架当前的最新版本是 Spring Framework 5 系列，在 Spring 4 的基础上增加了更多新特性，主要体现在以下几个方面。

(1) 反应式编程：基于 Reactor 3.1 构建的 Spring WebFlux 框架，支持 RxJava 2.1 并在 Tomcat、Jetty、Netty 或 Undertow 上运行。

(2) Java 8 和 Kotlin 的功能风格：整个框架中的 API 改进和 Kotlin 扩展，特别是对于 Bean 注册和功能性 Web 端点。

(3) 引入@Nullable 和@NotNull 注解来修饰可空的参数以及返回值，避免运行时导致 NPE(空指针)异常。

(4) 与 Java EE 8 API 集成：支持 Servlet 4.0、Bean Validation 2.0、JPA 2.2 以及 JSON

绑定 API(作为 Spring MVC 中 Jackson / Gson 的替代品)。

(5) JDK 9 支持：在运行时，类路径以及模块路径(目前基于文件名的“自动模块”)与 JDK 9 完全一致。

# 1.2 MyBatis

前面了解了 Spring 框架，它可以解决企业级应用程序开发的复杂性，降低各组件之间的耦合度，并且可以和第三方持久层框架进行良好的整合，下面我们就来学习持久层框架 MyBatis。

## 1.2.1 Hibernate 简介

在学习 MyBatis 之前，先来了解 Hibernate，因为它也是当今主流的 Java 持久层框架之一。

Hibernate 是由 Gavin King 于 2001 年创建的开放源代码的对象关系框架。它可以高效地构建具有关系对象持久性和查询服务的 Java 应用程序。Hibernate 是一个开放源代码的对象关系映射框架，它对 JDBC 进行了轻量级的对象封装，它将 POJO 与数据库表建立映射关系，是一个全自动的 ORM 框架，Hibernate 可以自动生成 SQL 语句，自动执行，并把开发人员从 95%的公共数据持续性编程工作中解放出来。

Hibernate 对 JDBC 进行了轻量级的对象封装，建立对象与数据库表的映射，是一个全自动的、完全面向对象的持久层框架。Hibernate 语言具有以下特点：

(1) 将对数据库的操作转换为对 Java 对象的操作，从而简化开发。通过修改一个“持久化”对象的属性，从而修改数据库表中对应的数据。

(2) 提供线程和进程两个级别的缓存，提升应用程序性能。

(3) 有丰富的映射方式，将 Java 对象之间的关系转换为数据库表之间的关系。

(4) 屏蔽不同数据库实现之间的差异。在 Hibernate 中只需要通过“方言”的形式指定当前使用的数据库，就可以根据底层数据库的实际情况，生成适合的 SQL 语句。

(5) 非侵入式：Hibernate 不要求持久化类实现任何接口或继承任何类，使用 POJO 即可。

## 1.2.2 MyBatis 简介

MyBatis 本是 Apache 的一个开源项目 iBatis，2010 年这个项目由 apache software foundation 迁移到了 google code，并且改名为 MyBatis。2013 年 11 月迁移到 Github。

iBatis 一词来源于 internet 和 abatis 的组合，是一个基于 Java 的持久层框架。iBatis 提供的持久层框架包括 SQL Maps 和 Data Access Objects(DAOs)。

MyBatis 是一款优秀的持久层框架，本教材采用的版本是 MyBatis 3.5.4，它支持定制化 SQL、存储过程以及高级映射。MyBatis 避免了几乎所有的 JDBC 代码和手动设置参数以及获取结果集。MyBatis 可以使用简单的 XML 或注解来配置和映射原生类型、接口和 Java 的 POJO 为数据库中的记录。

因为其具有封装少、映射多样化、支持存储过程、可以进行 SQL 优化等特点，使得它取代了 Hibernate 成为 Java 互联网中首选的持久层框架。但由于其需要手动编写 SQL 语句，需要编写 XML 文件，所以给开发人员也带来了一定的麻烦。当然 MyBatis 的增强工具(MyBatis-Plus)解决了这一问题，关于 MyBatis-Plus，建议大家在学完 MyBatis 后自行查阅资料学习。

## 1.2.3 Hibernate 和 MyBatis 的区别

Hibernate 是基于 ORM 思想的框架，它出身于 sf.net，现在已经成为 Jboss 的一部分。MyBatis 是另外一种优秀的 ORM 框架，目前属于 apache 的一个子项目。Hibernate 和 MyBatis 的区别主要体现在以下几个方面。

(1) 从开发上来比较，Hibernate 开发中，SQL 语句已经被封装，可以直接使用，加快了系统开发；MyBatis 属于半自动化，SQL 需要手工完成，稍微繁琐；因此程序员不用精通 SQL，只要懂得操作 POJO 就能够操作对应的数据库的表。

(2) 从 SQL 优化方面来比较，Hibernate 自动生成 SQL，有些语句较为繁琐，会多消耗一些性能；MyBatis 需要手动编写 SQL，可以避免不需要的查询，提高系统性能。

(3) 从对象管理上来比较，Hibernate 是完整的对象—关系映射的框架，开发工程中，无须过多关注底层实现，只要去管理对象即可；MyBatis 需要自行管理映射关系。

(4) 从缓存机制上来比较，两者具有相同点。Hibernate 和 MyBatis 的二级缓存除了采用系统默认的缓存机制外，都可以通过使用程序员自定义的缓存或使用其他第三方缓存方案，创建适配器来完全覆盖缓存行为。两者的不同点是，Hibernate 的二级缓存配置在 SessionFactory 生成的配置文件中进行详细配置，然后再在具体的表—对象映射中配置是哪种缓存。MyBatis 的二级缓存配置都是在每个具体的表—对象映射中进行详细配置，这样针对不同的表可以自定义不同的缓存机制，并且 MyBatis 可以在命名空间中共享相同的缓存配置和实例，通过 Cache-ref 来实现。

综上所述，我们可以发现 Hibernate 对于一些管理系统，如 ERP 系统的开发是十分便利的。因为对于管理系统而言，首先在于实现业务逻辑，然后才考虑性能，所以 Hibernate

在过去那个时代是主流持久框架。在当下移动互联网时代，SQL 的定制优化更符合移动互联网高并发、大数据、高性能、高响应的要求。因此 MyBatis 是当下主流的持久层框架，这也是本书中我们要讲解的重点。

## 1.3 Spring MVC 简介

Spring MVC 是 Spring 提供给 Web 应用的框架设计，是一种基于 Java 的实现了 Web MVC 设计模式的请求驱动类型的轻量级 Web 框架，通过实现 Model-View-Controller(MVC)模式很好地将数据、业务与视图进行分离。Spring MVC 和 Struts、Struts2 非常类似，但是 Struts 的兼容性并没有 Spring MVC 优秀，并且会存在类臃肿，漏洞较多易入侵。而 Spring MVC 结构层次清晰，类简单，与 Spring 框架实现无缝对接，已成为当今互联网时代的主流框架。

MVC 模型是一种架构型的模式，本身不引入新功能，只是帮助我们将开发的结构组织得更加合理，使展示与模型分离、流程控制逻辑、业务逻辑调用与展示逻辑分离，让这些元素之间达到松耦合的效果。

(1) Model (模型)封装了应用程序的数据和由它们组成的 POJO。

(2) View (视图)负责把模型数据渲染到视图上，将数据以特定的形式展现给用户。

(3) Controller (控制器)负责处理用户请求，并建立适当的模型把它传递给视图渲染。

在 Spring MVC 中还可以定义逻辑视图，通过其提供的视图解析器就能够很方便地找到对应的视图进行渲染，或者使用其消息转换的功能，比如在 Controller 方法内加入注解 @ResponseBody 后，Spring MVC 就可以通过其消息转换系统，将数据转换为 JSON，提供给前端使用。

## 单元小结

- 容器在 J2EE 架构中占有很重要的地位，当前流行的是 EJB 容器和“轻量级”容器。EJB 容器有很强的侵入性，而“轻量级”容器可以管理 POJO 对象。
- Spring 的核心是控制反转(IoC)和面向切面(AOP)。
- Spring 也是一个“一站式”的开发框架，覆盖了 J2EE 开发中的各个分层，可以和多个子框架和开源项目整合，适用于多种应用场景。
- MyBatis 是一款优秀的持久层框架，它支持定制化 SQL、存储过程以及高级映射。

## 单元自测

1. 下列对于企业级应用的描述，错误的是(　　)。
   A. 企业级应用中的系统可能分布在不同的地方
   B. 企业级应用对安全性要求很高
   C. 企业级应用系统彼此间是相互独立的，不需要和现有的系统整合
   D. 企业级应用中数据的一致性要求很高，所以必须有严格的事务性
2. 在 J2EE 架构的组成部分中，不包括(　　)。
   A. 业务层
   B. 网络层
   C. 表示层
   D. 企业信息层
3. 下列功能是容器应该提供的，但(　　)除外。
   A. 程序对象的生命周期管理
   B. 声明式事务、安全等企业级服务
   C. 与现有开发框架的整合
   D. 对程序中业务对象的依赖关系的管理
4. 关于 EJB 容器和轻量级容器，下列说法正确的是(　　)。
   A. 轻量级容器启动速度更快
   B. 轻量级容器使用的是无侵入性的 POJO 对象，而 EJB 容器中的业务对象必须使用特定的 API
   C. 基于轻量级容器的代码的可测性更强
   D. 只能使用两者中的一个进行程序的开发
5. 关于 IoC 的理解，下列描述正确的两项是(　　)。
   A. 控制反转
   B. 对象被动地接受依赖类
   C. 对象主动去找依赖
   D. 一定要用接口

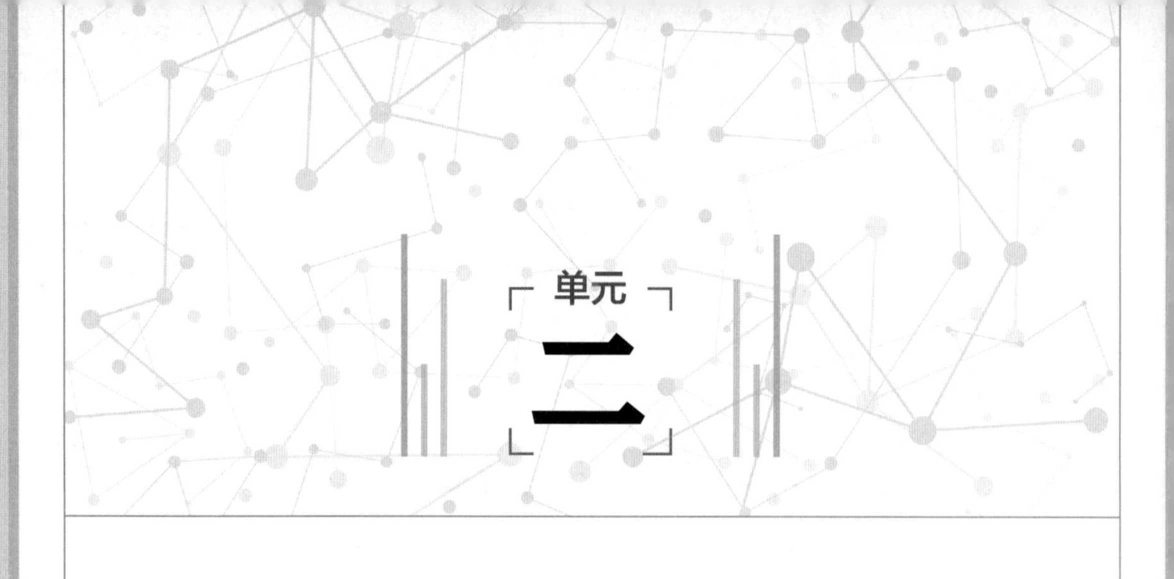


# MyBatis核心组件

## 课程目标

❖ 了解 MyBatis 的特点

❖ 了解 MyBatis 的核心构成

❖ 掌握 SqlSessionFactory 的使用

❖ 了解映射器的执行原理

❖ 掌握 MyBatis 生命周期

**简介**

MyBatis 作为主流的持久层框架之一，在操作数据库上具有很大的便捷性，不屏蔽 SQL，便于 SQL 语句的优化，提高系统性能。MyBatis 避免了几乎所有的 JDBC 代码和手动设置参数以及获取结果集的操作，它提供强大、灵活的映射机制，并且拥有功能丰富的组件。

本章将围绕 MyBaits 核心组件展开讲解。

## 2.1 环境准备

先来创建 Maven 项目。Maven 是一个优秀的项目构建和管理工具，后面的内容都会在 Maven 构建的项目基础上进行讲解和测试，本书使用的开发工具是目前比较流行且深受开发者欢迎的 IntelliJ IDEA，本书后面统一简称为 IDEA。

先在 IDEA 中创建一个基本的 Maven 项目，按照如下步骤操作即可。

(1) 打开 IDEA 开发工具，在菜单栏中找到【窗口】，选择【File】→【New】→【Project】命令，打开【New Project】对话框，在左侧列表中选择【Maven】并选中【Create from archetype】复选框，如图 2-1 所示。

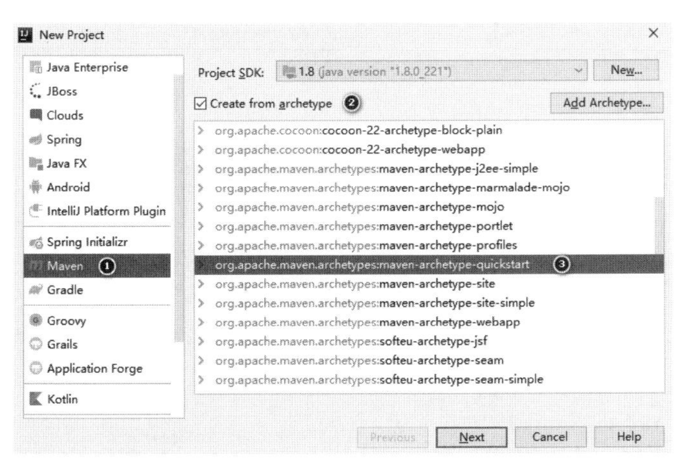


图 2-1

(2) 选中模板 maven-archetype-quickstart，单击【Next】按钮，接下来填写项目信息。在 GroupId 文本框中填写公司或组织域名；在 ArtifactId 文本框中填写项目名；在 Version 文本框中填写项目版本号。填写完成后的界面如图 2-2 所示。

图 2-2

(3) 完成以上操作，单击【Next】按钮，接下来选择 Maven 地址、maven settings.xml 配置文件、repository 本地仓库(需提前配置好 Maven 环境)，最后单击【Finish】按钮，如图 2-3 所示。

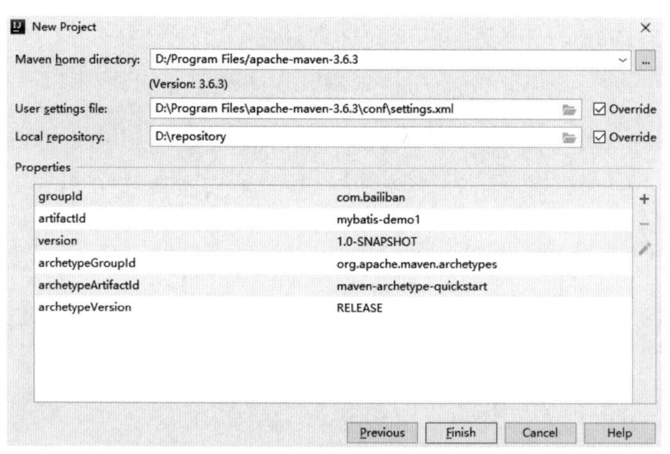


图 2-3

(4) 在 pom 文件中加入以下 MyBatis 依赖坐标。

```
<dependency>
    <groupId>org.mybatis</groupId>
    <artifactId>mybatis</artifactId>
    <version>3.5.4</version>
</dependency>
```

至此我们的项目准备工作就完成了。

## 2.2 ORM 和 MyBatis 的特点

ORM 即 Object Relational Mapping(对象—关系映射)的简称。项目中的业务实体有两种表现形式：对象和关系数据，即在内存中表现为对象，在数据库中表现为关系数据。

ORM 框架是对象关系映射，因为对象之间可以存在关联和继承关系，但是在数据库中，关系数据无法表达多对多关联和继承关系。因此，要想完成对象和关系型数据库之间的数据交互，就需要建立对象和关系(业务实体的两种表现形式)的映射，只要提供了持久化类与表的映射关系，ORM 框架在运行时就能参照映射文件的信息，把对象持久化到数据库中。ORM 框架一般以中间件的形式存在，如图 2-4 所示。

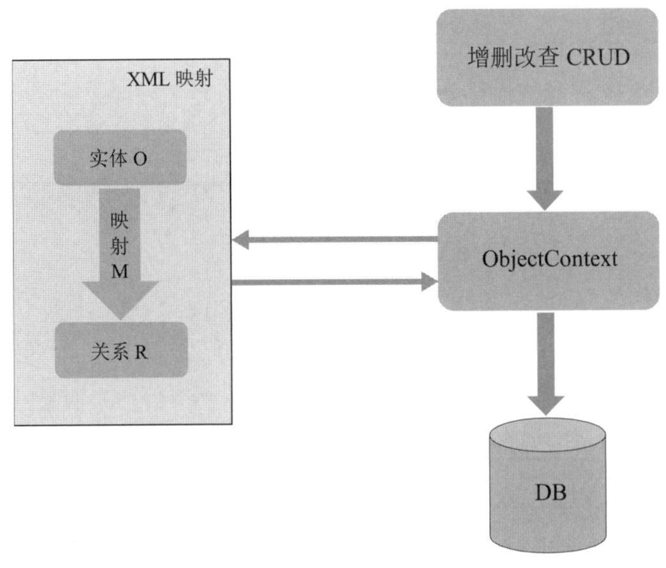


图 2-4

从系统结构上看，采用 ORM 的系统一般都是多层系统，系统的层次多了，效率就会降低。ORM 是一种完全面向对象的做法，而面向对象的做法也会对性能产生一定的影响，因此一个半自动的 ORM 框架 MyBatis 应运而生。

MyBatis 是一个 Java 持久层框架，它封装少、高性能、可优化、维护简单等优点成为了目前 Java 移动互联网网站服务的首选持久层框架，它特别适合分布式和大数据网络数据库编程。之所以称它为半自动，是因为它需要手工匹配提供 POJO、SQL 和映射关系，而全表映射的 Hibernate 只需要提供 POJO 和映射关系便可，如图 2-5 所示。

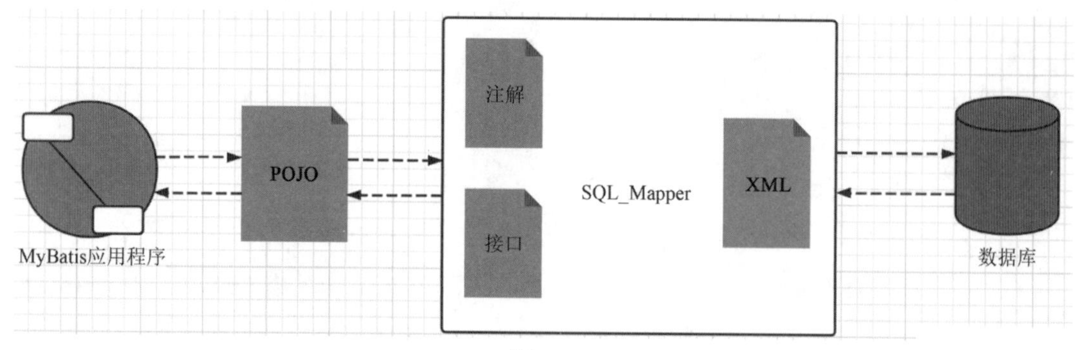


图 2-5

## 2.3 MyBatis 的基本构成

学习 MyBatis 框架，首要任务就是学习 MyBatis 的基本构成，即 MyBatis 核心组件。MyBatis 核心组件包括四部分，分别为：

(1) SqlSessionFactoryBuilder( 构 造 器 )： 它 会 根 据 配 置 或 者 代 码 来 生 成 SqlSessionFactory，采用的是分步构建的 Builder 模式。

(2) SqlSessionFactory (工厂接口)：依靠它来生成 SqlSession，使用的是工厂模式。

(3) SqlSession(会话)：一个既可以发送 SQL 执行返回结果，又可以获取 Mapper 的接口。在现有的技术中，一般会让其在业务逻辑代码中“消失”，而使用的是 MyBatis 提供的 SQL Mapper 接口编程技术，它能提高代码的可读性和可维护性。

(4) SQL Mapper(映射器)：MyBatis 新设计存在的组件，它由一个 Java 接口和一个 XML 文件(或注解)构成，需要给出对应的 SQL 和映射规则，它可以发送 SQL 并返回结果。

下面用图解的形式来展示 MyBatis 核心组件之间的关系，如图 2-6 所示。

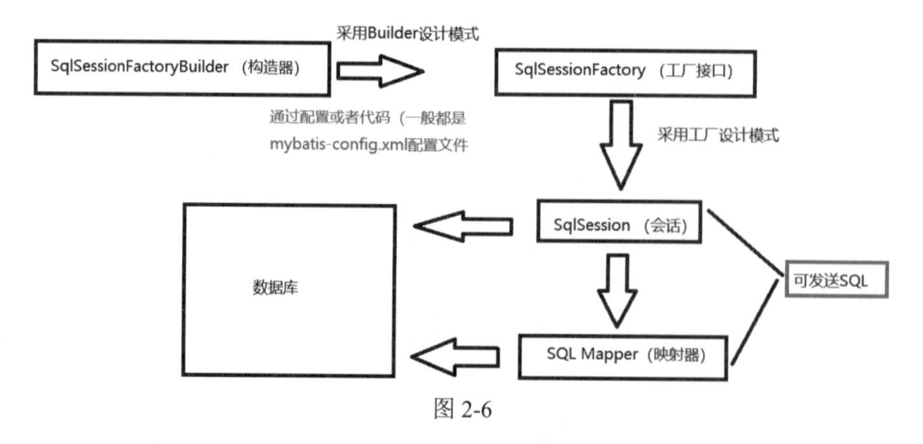


图 2-6

注意，无论是映射器还是 SqlSession 都可以发送 SQL 到数据库执行，下面我们一一学习这些组件的用法。

## 2.4 SqlSessionFactoryBuilder 组件

SqlSessionFactoryBuilder 的作用在于依据配置信息或者代码生成 SqlSessionFactory。而 MyBatis 提供了构造器 SqlSessionFactoryBuilder。查看 MyBatis 源码，可以看到该类有一个默认构造器和多个 build()的重载方法，如图 2-7 所示。

SqlSession
SqlSessionException
SqlSessionFactory
SqlSessionFactoryBuilder
SqlSessionFactoryBuilder()
build(Configuration):SqlSessionFactory
build(InputStream):SqlSessionFactory
build(InputStream, Properties):SqlSessionFactory
build(InputStream, String):SqlSessionFactory
build(InputStream, String, Properties):SqlSessionFactory
build(Reader):SqlSessionFactory
build(Reader, Properties):SqlSessionFactory
build(Reader, String):SqlSessionFactory
build(Reader, String, Properties):SqlSessionFactory
SqlSessionManager
TransactionIsolationLevel
org.apache.ibatis.session.defaults
org.apache.ibatis.transaction

图 2-7

进一步分析便可发现，该类中有三个主要的方法：

```
public SqlSessionFactory build(Reader reader, String environment, Properties properties){...};
public SqlSessionFactory build(InputStream inputStream, String environment, Properties properties){...};
public SqlSessionFactory build(Configuration config){...}
```

第一个方法通过字符流读取配置文件信息；第二个方法通过字节流读取配置文件信息；第三个方法通过代码的方式创建 SqlSessionFactory。通过字符流或字节流读取配置文件信息创建 SqlSessionFactory，实际均是利用第三个方法，最终都会创建 Configuration 对象以创建 SqlSessionFactory，其本质是一致的。

## 2.5 SqlSessionFactory 组件

SqlSessionFactory 用于生产 SqlSession 会话。由前文可知，创建 SqlSessionFactory 有两种方式，一种是使用 XML 配置文件的方式，另一种则是使用 Java 代码的方式。以下举例简要介绍用两种方式创建 SqlSessionFactory 的过程。

## 2.5.1 使用 XML 方式构建 SqlSessionFactory

使用 XML 构建 SqlSessionFactory 是开发中经常使用的一种方式，XML 配置可以让软件更具有扩展性，具体操作步骤如下：

(1) 创建用于存放资源文件的目录。鼠标右键单击【src】选项，然后选择【New】→【Directory】命令，此时会弹出如图 2-8 所示的界面。分别双击 main\resources 和 test\resources，创建主资源文件夹和测试资源文件夹。

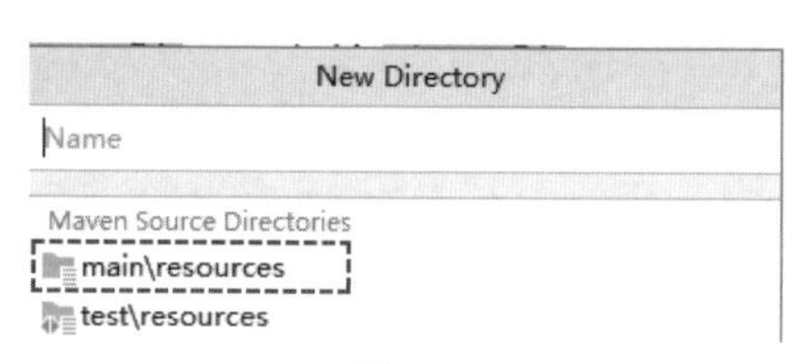


图 2-8

(2) 在 pom 文件中除去配置 MyBatis 和 Junit 的依赖外，还需要加入 MySQL 连接的依赖坐标完成 MySQL 数据库的操作，加入日志依赖坐标以方便查看系统操作的日志信息，加入 lombok 插件依赖坐标可以精简 POJO 实体类的编写。相关配置代码如下：

```
<dependency>
        <groupId>mysql</groupId>
        <artifactId>mysql-connector-java</artifactId>
        <version>8.0.15</version>
    </dependency>
    <dependency>
        <groupId>org.projectlombok</groupId>
        <artifactId>lombok</artifactId>
        <version>1.18.10</version>
    </dependency>
    <dependency>
        <groupId>log4j</groupId>
        <artifactId>log4j</artifactId>
        <version>1.2.17</version>
</dependency>
```

(3) 在 src/main/java 目录下创建 User 实体类，并放到 com.bailiban.pojo 包下，代码如下：

```
@Data
public class User {
    private int id;
    private String username;
```

```
    private Date birthday;
    private String address;
    private String sex;
}
```

(4) 在 src/main/resources 目录下创建 db.properties 文件，内容如下：

```
jdbc.driver=com.mysql.cj.jdbc.Driver
jdbc.url=jdbc:mysql://localhost:3306/mybatis?serverTimezone=GMT%2b8&useSSL=false&useUnicode
      =true&characterEncoding=utf-8
jdbc.username=root
jdbc.password=123456
```

(5) 在 src/main/resources 目录下创建 mybatis-config.xml 配置文件，文件内容如下：

```
<?xml version="1.0" encoding="UTF-8"?>
<!DOCTYPE configuration PUBLIC "-//mybatis.org//DTD Config 3.0//EN"
"http://mybatis.org/dtd/mybatis-3-config.dtd">
<configuration>

    <!--引入外部数据库连接配置文件-->
    <properties resource="db.properties"/>

    <!--配置 MyBatis 运行环境-->
    <environments default="development">
        <environment id="development">
            <!-- type="JDBC" 代表使用 JDBC 的提交和回滚来管理事务 -->
            <transactionManager type="JDBC"/>
            <dataSource type="POOLED">
                <property name="driver" value="${jdbc.driver}"/>
                <property name="url" value="${jdbc.url}"/>
                <property name="username" value="${jdbc.username}"/>
                <property name="password" value="${jdbc.password}"/>
            </dataSource>
        </environment>
    </environments>
    <mappers>
        <!--映射文件-->
        <mapper resource="com/bailiban/mapper/UserMapper.xml"/>
    </mappers>
</configuration>
```

对配置文件的简要说明如下。

- <properties>：配置 properties，db.properties 文件保存数据库连接信息。
- <environment>：元素的定义，这里描述的是数据库。它里面的<transactionManager>元素是配置事务管理器，这里采用的是 MyBatis 的 JDBC 管理器方式。然后采用

<dataSource>元素配置数据库，其中属性 type＝"POOLED"代表采用 MyBatis 内部提供的连接池方式，最后定义一些关于 JDBC 属性的信息。

- 配置映射器：引入一个 XML(UserMapper.xml)文件，它提供了 SQL 和 POJO 的映射规则，包含映射器中的信息，关于映射器我们会在后面进行讲解。

(6) 完成以上步骤后，就开始使用 XML 配置方式创建 SqlSessionFactory。代码如下：

```
public class SqlSessionFactoryWithXmlUtil {

    /**
     * 获取当前对象用于加锁
     */
    private static final Class<SqlSessionFactoryWithXmlUtil> CLASS_LOCK =
                SqlSessionFactoryWithXmlUtil.class;
    /**
     * 首先创建静态成员变量 SqlSessionFactory，静态变量被所有的对象共享。
     */
    private static SqlSessionFactory sqlSessionFactory = null;

    /**
     * 构造私有函数
     */
    private SqlSessionFactoryWithXmlUtil() {
    }

    /**
     * 单例模式
     *
     * @return SqlSessionFactory
     */
    public static SqlSessionFactory getSqlSessionFactory() {
        synchronized (CLASS_LOCK) {
            if (sqlSessionFactory == null) {
                // 1.创建构建者对象 builder
                SqlSessionFactoryBuilder builder = new SqlSessionFactoryBuilder();
                // 2.读取 MyBatis 配置文件
                InputStream input = null;
                try {
                    input = Resources.getResourceAsStream("mybatis-config.xml");
                } catch (IOException e) {
                    e.printStackTrace();
                }
                // 3.使用构建者创建工厂对象 SqlSessionFactory
                sqlSessionFactory = builder.build(input);
            }
            return sqlSessionFactory;
        }
```

```
    }
}
```

创建 SqlSessionFactory 实际分为三个步骤，首先创建 mybatis-config.xml 配置文件，然后通过 IO 流读取配置文件的信息，最后使用 SqlSessionFactoryBuilder 的 build 方法完成 SqlSessionFactory 的创建。流程十分简单，因为这里使用了构建者(Builder)模式，其中大量的 XpathParser 解析、配置或语法的解析、反射生成对象、存入结果缓存等细节对开发者都进行了隐藏。

使用 XML 方式构建 SqlSessionFactory，信息在配置文件中，并和 Java 代码进行了解耦，有利于后续的维护和修改。本书推荐这种方式。

## 2.5.2 使用 Java 代码方式构建 SqlSessionFactory

在上节中，我们已经演示了使用 XML 方式来构建 SqlSessionFactory，这里为了拓宽学习面，也展示使用 Java 代码方式构建 SqlSessionFactory 的方法，便于大家进行对比。代码如下：

```
public class SqlSessionFactoryWithCodeUtil {
    /**
     * 获取当前对象用于加锁
     */
    private static final Class<SqlSessionFactoryWithCodeUtil> CLASS_LOCK =
                SqlSessionFactoryWithCodeUtil.class;
    /**
     * 首先创建静态成员变量 sqlSessionFactory，静态变量被所有的对象共享。
     */
    private static SqlSessionFactory sqlSessionFactory = null;

    /**
     * 构造私有函数
     */
    private SqlSessionFactoryWithCodeUtil() {
    }

    /**
     * 单例模式
     *
     * @return SqlSessionFactory
     */
    public static SqlSessionFactory getSqlSessionFactory() {
        synchronized (CLASS_LOCK) {
            if (sqlSessionFactory == null) {
```

```
                init();
            }
            return sqlSessionFactory;
        }
    }
    /**
     * 初始化 SqlSessionFactory
     */
    private static void init() {
        // 1. 创建连接池
        PooledDataSource dataSource = new PooledDataSource();
        // 2. 设置连接信息
        dataSource.setDriver("com.mysql.cj.jdbc.Driver");
        dataSource.setUrl("jdbc:mysql://localhost:3306/mybatis?serverTimezone
        =GMT%2b8&useSSL=false&useUnicode=true&characterEncoding=utf-8");
        dataSource.setUsername("root");
        dataSource.setPassword("123456");
        // 3. 事务管理方式
        TransactionFactory transaction = new JdbcTransactionFactory();
        // 4. 创建运行环境
        Environment environment = new Environment("development", transaction, dataSource);
        // 5. 创建 Configuration 对象
        Configuration configuration = new Configuration(environment);
        // 6. 添加映射器
        configuration.addMapper(UserDao.class);
        // 7. 构建 SqlSessionFactory
        sqlSessionFactory = new SqlSessionFactoryBuilder().build(configuration);
    }
}
```

和 2.2.1 节中的代码比较，我们会发现和 XML 配置方式比较类似，只是这里不再是读取 mybatis-config.xml 配置文件，而是以 Java 代码的方式编写配置信息。这种方式导致配置和代码混在一起，代码冗长，不便于维护修改。

## 2.6 MyBatis 核心接口 SqlSession

SqlSession是MyBatis的核心接口，主要进行持久化操作。在MyBatis中有两个实现类，DefaultSqlSession和SqlSessionManager。DefaultSqlSession是单线程使用的，而 SqlSessionManager在多线程环境下使用。SqlSession的作用类似于一个JDBC中的Connection对象，代表着一个连接资源的启用。SqlSession提供的方法如图2-9所示。

图 2-9

由图 2-9 可知，一方面，SqlSession 提供了进行增删改查的方法，同时可进行事务提交或者回滚；另一方面，提供了获取映射器的方法。需要注意：SqlSession 的实例是线程不安全的，不能将其放在类的静态字段或者实例字段中；使用完 SqlSession 之后，应该利用 finally 块来确保关闭它。

综上所述，我们可以知道 SqlSession 的作用主要有以下三个：

- 获取 Mapper 接口。
- 发送 SQL 给数据库。
- 控制数据库事务。

下面通过 SqlSessionFactory 来创建 SqlSession，并查看 SqlSession 在操作事务时的伪代码：

```
//定义 SqlSession
SqlSession sqlSession = null;
try{
    //打开 SqlSession 会话
    sqlSession = SqlSessionFactory.openSession();
    sqlSession.commit();//提交事务
}catch (Exception ex) {
    sqlSession.rollback();//回滚事务
}finally {
    //在 finally 语句中确保资源被顺利关闭
    if (sqlSession != null) {
```

```
        sqlSession.close();
    }
}
```

执行完数据库操作后，可以使用 SqlSession 的 commit 方法提交事务，如果出现异常则使用 rollback 方法进行事务回滚。

注意：使用数据库的连接资源后要及时关闭它，如果不及时关闭，数据库的连接资源就会很快被消耗光，整个系统将陷入瘫痪状态，所以这里使用 finally 语句使其顺利关闭。

有了 SqlSession 对象后，接下来需要实现映射器的功能，然后就可以获取 Mapper 接口和发送 SQL，下面我们就来讲解映射器。

# 2.7 映射器

映射器是 MyBatis 中最核心的组件之一，MyBatis 3 之前的版本只支持 XML 映射器，即所有的 SQL 语句都必须在 XML 文件中配置。而从 MyBatis 3 开始可以支持接口映射器，这种映射器方式允许以 Java 代码的方式注解定义 SQL 语句，非常简洁。下面我们逐一详细讲解。

## 2.7.1 用 XML 实现映射器

用 XML 文件定义映射器分为两个部分：定义映射器接口、定义 XML 映射器。

(1) 定义映射器接口。在 src/main/java 目录下创建 UserDao 接口，并放到 com.bailiban.dao 包下，代码如下：

```
public interface UserDao {
    /**
     * 根据用户 id 查询用户
     * @param id 用户 id
     * @return User
     */
    User findUserById(@Param("id") Integer id);

    /**
     * 查询所有
     * @return List<User>
     */
    List<User> selectAll();
}
```

(2) 定义 XML 映射器。在 src/main/resources 目录下创建 UserMapper.xml 映射文件，并放在 com.bailiban.mapper 三级目录下，需要注意的是，在创建 com.bailiban.mapper 目录时，需要一级一级地创建，即先创建 com 目录，再创建 bailiban 目录，最后创建 mapper 目录。UserMapper.xml 内容如下：

```
<?xml version="1.0" encoding="UTF-8" ?>
<!DOCTYPE mapper
        PUBLIC "-//mybatis.org//DTD Mapper 3.0//EN"
        "http://mybatis.org/dtd/mybatis-3-mapper.dtd">
<mapper namespace="com.bailiban.dao.UserDao">
    <select id="findUserById" parameterType="integer" resultType="com.bailiban.pojo.User">
        select *
        from tab_user
        where id = #{id}
    </select>
    <select id="selectAll" resultType="com.bailiban.pojo.User">
        select *
        from tab_user
    </select>
</mapper>
```

(3) 配置 XML 映射器。在 2.5.1 节使用 XML 方式构建 SqlSessionFactory 的过程中，创建了 mybatis-config.xml 文件，其中有如下一段配置代码：

```
<mappers>
        <!-- 映射文件 -->
        <mapper resource="com/bailiban/mapper/UserMapper.xml"/>
</mappers>
```

它的作用就是注册上面定义好的XML映射器mapper文件，MyBatis会扫描这些mapper文件的信息，完成数据映射和 SQL 的执行，这样就完成了使用 XML 方式来定义映射器。

UserMapper.xml 映射器文件结构也比较简单，大致讲解如下：

- <mapper>元素中的属性 namespace 所对应的是一个映射器接口的全限定名，于是 MyBatis 上下文就可以通过它找到对应的接口。
- <select>元素描述的是查询语句，属性 id 标识了这条 SQL 语句，并且映射器接口中的方法名和 id 属性值是相匹配的。parameterType 属性指定传递给 SQL 的参数类型，resultType 属性指明返回结果类型，当返回类型为对象时，使用全限定名或别名(使用别名需要在 mybatis-config.xml 中进行配置)。
- #{id}作为参数的占位符，表示传递进来的参数 id。

在以上配置中，我们发现并没有配置 SQL 执行后和 user 的对应关系，那么查询结果列和实体类属性是如何映射的呢？其实这里 MyBatis 在默认情况下提供自动映射，只要

SQL 返回的列名和 POJO 对应起来即可。当然如果自动映射满足不了需求，此时可以通过 resultMap 来映射，其内容会在后面章节进行详解。

### 2.7.2 用注解实现映射器

使用注解方式定义映射器就比较简单了，只需要定义一个映射器接口即可，然后通过注解的方式来注入 SQL。代码如下：

```
public interface UserDao {
    /**
     * 根据用户 id 查询用户
     *
     * @param id  用户 id
     * @return User
     */
    @Select("select * from tab_user where id=#{id}")
    User findUserById(@Param("id") Integer id);

    /**
     * 查询所有
     *
     * @return List<User>
     */
    @Select("select * from tab_user")
    List<User> selectAll();
}
```

使用注解方式实现映射器，在处理一些比较简单的 SQL 语句时，使用起来非常简洁。但是实际工作中 SQL 语句会比我们这里的举例复杂得多，而使用 XML 方式能更好地处理相对复杂的 SQL 结构，同时 MyBatis 官方推荐使用的也是 XML 方式，因此本书以 XML 方式为主讨论 MyBatis 的应用。另外需要注意的是同一个映射器接口方法上不能同时使用 XML 方式和注解方式实现映射器。

### 2.7.3 用 SqlSession 发送 SQL

完成映射器的定义后，就可以通过 SqlSession 发送 SQL 了，代码如下：

```
@Test
public void selectOne2() {
    SqlSession sqlSession = null;
    try {
        SqlSessionFactory sqlSessionFactory = SqlSessionFactoryWithXmlUtil.getSqlSessionFactory();
        // 生成 sqlSession
```

```
            sqlSession = sqlSessionFactory.openSession();
            //使用 sqlSession 的 selectOne 方法查询
            User user = sqlSession.selectOne("com.bailiban.dao.UserDao.findUserById", 41);
            System.out.println(user);
        } finally {
            if (sqlSession != null) {
                sqlSession.close();
            }
        }
    }
```

代码中的 selectOne 方法表示的是执行查询且返回一个对象。方法中包含一个 String 类型的参数和一个 Object 类型的参数。

其中 String 类型的参数是由 namespace 命名空间加上 SQL id 组合而成的。这样 MyBatis 就会找到对应的 SQL。如果在 MyBatis 中只有一个 id 为 findUserById 的 SQL，那么 String 类型参数的全限定名也可以省略。如下所示：

```
User user = sqlSession.selectOne("findUserById", 41);
```

## 2.7.4　用 Mapper 接口发送 SQL

SqlSession 除了可以直接发送 SQL 外，还可以获取 Mapper 接口，通过 Mapper 接口发送 SQL，代码如下：

```
@Test
public void selectOne() {
    SqlSession session = null;
    try {
        SqlSessionFactory sqlSessionFactory = SqlSessionFactoryWithXmlUtil.getSqlSessionFactory();
        session = sqlSessionFactory.openSession();
        // 1.使用 SqlSession 创建 dao 接口的代理对象
        UserDao dao = session.getMapper(UserDao.class);
        // 2.使用代理对象执行查询
        User user = dao.findUserById(41);
        System.out.println(user);
    } finally {
        if (session != null) {
            session.close();
        }
    }
}
```

以上代码展示了使用 SqlSession 调用 getMapper 方法获取映射接口 UserDao，实际我们更习惯称之为 UserMapper，这里只是为了延续大家之前对 Dao 层的叫法，因此创建的映射接口名叫 UserDao。在后面的章节中我们会改正过来。

在这里会发现，最终是通过 Mapper 接口，即 Dao 层的 UserDao 来调用接口方法进行 SQL 的操作。有的学员可能会比较疑惑，我们在这里并没有去编写接口的实现类，但能正常运行，是因为 MyBatis 利用动态代理帮我们生成了接口的实现类，这个类就是 org.apache.ibatis.binding.MapperProxy。关于动态代理，大家可以自己查阅资料进行了解，这里不作详细介绍。

### 2.7.5 对比两种 SQL 发送方式

在 2.6.3 和 2.6.4 节中展示了 MyBatis 存在的两种发送 SQL 的方式，这里推荐使用 Mapper 接口方式发送 SQL。理由如下：

- 使用 Mapper 接口编程更符合面向对象的思想，通过调用自己的业务持久层接口，完成对数据的操作，它能提高代码的可读性和可维护性。而 SqlSession 直接发送 SQL 方式需要调用其自身的方法，然后需要一个 SQL id 来匹配 SQL，比较难以理解。
- 由于 SqlSession 调用的是 SQL id，其类型为 String 类型，所以 IDEA 无法进行错误提示。例如，findUserById 写成了 findUerByd，如下代码所示，少写了一个 I，此时 IDEA 只有在运行时才会报错。而使用 Mapper 接口方式，IDEA 会提示错误和建议。

```
User user = sqlSession.selectOne("findUserByd", 41);
```

目前使用 Mapper 接口编程已经成为主流方式，尤其在和 Spring 整合后，Mapper 接口的方式更加简单，所以本书中后面的代码演示主要以 Mapper 接口方式进行。

## 2.8 生命周期

前面已经学习了 MyBatis 组件的创建和基本应用，但如果不了解其生命周期，错误的使用会导致非常严重的并发问题。因此有必要掌握 MyBatis 组件的生命周期。

所谓的生命周期，就是每一个对象应该存活的时间，比如一些对象一次用完后就要关闭，使它们被 Java 虚拟机(JVM)销毁，以避免继续占用资源，所以我们会根据每一个组件的作用去确定其生命周期。

### 2.8.1　SqlSessionFactoryBuilder

在 2.4 节中我们已经讲到 SqlSessionFactoryBuilder 的作用在于依据配置信息或者代码生成 SqlSessionFactory，它的作用就是一个构建器，一旦构建了 SqlSessionFactory，它的作用就已经完结，这个类就不需要存在了，所以可以把它作为一个局部变量使用。

### 2.8.2　SqlSessionFactory

SqlSessionFactory 的作用是创建 SqlSession，而 SqlSession 就是一个会话，相当于 JDBC 中的 Connection 对象。因此可以把 SqlSessionFactory 想象为数据库连接池。SqlSessionFactory 一旦被创建就应该在应用的运行期间一直存在，所以它的生命周期就是 MyBatis 的应用周期。而且为了避免连接资源被耗尽，会将 SqlSessionFactory 作为一个单例，让它在应用中共享。

### 2.8.3　SqlSession

每个线程都有它自己的 SqlSession 实例，SqlSession 实例不能被共享，也不是线程安全的。上文提到 SqlSession 就相当于一个数据库连接(Connection 对象)，可以在一个事务里面执行多条 SQL，然后通过它的 commit、rollback 等方法，提交或者回滚事务。所以它应该存活在一个业务请求中，处理完整个请求后，应该关闭这条连接，让它归还给 SqlSessionFactory，否则数据库资源就很快被耗费精光，系统就会瘫痪，所以使用 try...catch...finally...语句来保证其正确关闭。

### 2.8.4　Mapper

Mapper 是一个接口，没有实现类，它的作用是发送 SQL 然后返回我们需要的结果，或者执行 SQL 从而修改数据库的数据，所以它应该在一个请求中，一旦处理完了相关的业务，就应该废弃它。

同时 Mapper 接口的实例是从 SqlSession 中获得的，所以它的最大生命周期至多和 SqlSession 保持一致，但是由于 SqlSession 的关闭，它的数据库连接资源也会消失，所以它的生命周期应该小于或等于 SqlSession 的生命周期。

## 单元小结

- MyBatis 是一个半自动的 ORM 框架。
- MyBatis 核心组件包括 SqlSessionFactoryBuilder、SqlSessionFactory、SqlSession、Mapper。
- SqlSessionFactory 使用单例模式。
- SqlSession 即用即创建，用完要关闭。

1. 关于 MyBatis 的描述，错误的是(　　)。
   A. MyBatis 是 JDBC 的轻量级封装，把 Sql 和 Java 代码独立出来
   B. MyBatis 里面的核心处理类叫做 SqlSession
   C. MyBatis 是一个半 ORM(对象关系映射)框架
   D. 使用${}可以有效防止 SQL 注入
2. 以下关于 MyBatis 核心类生命周期的说法，错误的是(　　)。
   A. SqlSessionFactoryBuilder 实例的最佳作用域是方法范围，也就是定义为本地方法变量
   B. SqlSessionFactory 实例的生命周期应该在整个应用的执行期间都存在
   C. SqlSession 实例是线程不安全的，因此其生命周期应该是请求或方法范围
   D. SqlSession 实例通常定义为一个类的静态变量
3. MyBatis 核心组件包括(　　)个部分。
   A. 2　　　　B. 3
   C. 4　　　　D. 以上都不对
4. SqlSession 有两个实现类，其中在多线程环境下使用的是(　　)。
   A. DefaultSqlSession　　　　B. SqlSessionManager
   C. SqlSessionFactory　　　　D. SqlSessionFactoryBuilder
5. 下列关于 MyBatis 优点描述，不正确的是(　　)。
   A. SQL 写在 XML 里，便于统一管理和优化
   B. 提供映射标签，支持对象与数据库的 ORM 字段关系映射
   C. 提供 XML 标签，支持编写动态 SQL
   D. 数据库移植性好，可以随意更换数据库

## 上机实战

### 上机目标

- 掌握对 MyBatis 组件的应用。
- 完成数据库的简单查询操作。

### 上机练习

**◆ 第一阶段◆**

#### 练习 1：使用 XML 实现映射器方式完成查询

**【问题描述】**

使用 XML 映射器方式根据用户的名称查询用户信息。

**【问题分析】**

根据文中案例，使用<select>标签编写 UserMapper.xml 映射文件，传递一个 name 参数完成查询操作。

**【参考步骤】**

(1) 创建 Maven 应用，配置好 MyBatis 环境。

(2) 创建数据库 mybatis，新建表 tab_user。建表语句如下：

```
CREATE TABLE `tab_user` (
  `id` int(11) NOT NULL AUTO_INCREMENT,
  `username` varchar(32) NOT NULL COMMENT '用户名称',
  `birthday` date DEFAULT NULL COMMENT '生日',
  `sex` char(1) DEFAULT NULL COMMENT '性别',
  `address` varchar(256) DEFAULT NULL COMMENT '地址',
  PRIMARY KEY (`id`)
) ENGINE=InnoDB AUTO_INCREMENT=48 DEFAULT CHARSET=utf8;
```

(3) 创建 User 实体类，代码见 2.5.1 节。

(4) 创建 db.properties，代码见 2.5.1 节。

(5) 在 src/main/resources 下创建 log4j.properties，代码如下所示。

```
#全局配置
log4j.rootLogger=DEBUG,stdout
#日志配置
log4j.logger.org.mybatis=DEBUG
#控制台输出配置
log4j.appender.stdout=org.apache.log4j.ConsoleAppender
```

```
log4j.appender.stdout.layout=org.apache.log4j.PatternLayout
log4j.appender.stdout.layout.ConversionPattern=%d [%t] %-5p [%c] - %m%n
```

这里将日志输出级别设置成DEBUG级别，这样就可以把最详细的日志信息显示出来，方便开发者调试，在生产环境中，可以设置为INFO级别。

(6) 创建mybatis-config.xml配置文件，代码见2.5.1节。

(7) 在com.bailiban.mapper包下创建UserMapper接口，添加根据姓名查询用户的接口方法。代码如下所示。

```
public interface UserMapper {
    /**
     * 根据姓名查询User
     * @param name：用户名
     * @return User
     */
    User findUserByName(String name);
}
```

(8) 在src/main/resources下的com.bailiban.mapper包下创建UserMapper.xml映射器文件，代码如下所示。

```
<?xml version="1.0" encoding="UTF-8" ?>
<!DOCTYPE mapper
        PUBLIC "-//mybatis.org//DTD Mapper 3.0//EN"
        "http://mybatis.org/dtd/mybatis-3-mapper.dtd">
<mapper namespace="com.bailiban.mapper.UserMapper">
    <select id="findUserByName" parameterType="string" resultType="com.bailiban.pojo.User">
        select *
        from tab_user
        where username = #{name}
    </select>
</mapper>
```

(9) 在测试包下创建测试类，使用Junit测试单元进行测试，代码如下所示。

```
@Test
public void findUserByName() {
    SqlSession sqlSession = null;
    try {
        SqlSessionFactory sqlSessionFactory = SqlSessionFactoryWithXmlUtil.getSqlSessionFactory();
        // 生成sqlSession
        sqlSession = sqlSessionFactory.openSession();
        // 使用sqlSession的selectOne方法查询
        UserMapper mapper = sqlSession.getMapper(UserMapper.class);
        User user = mapper.findUserByName("百里半");
        System.out.println(user);
    } finally {
```

```
            if (sqlSession != null) {
                sqlSession.close();
            }
        }
    }
```

代码中的 SqlSessionFactoryWithXmlUtil 工具类请参照 2.5.1 节中使用 xml 配置方式创建 SqlSessionFactory。

(10) 运行测试代码，控制台显示出如下信息，如图 2-10 所示。

```
✓ Tests passed: 1 of 1 test – 2 s 900 ms
"D:\Program Files\Java\jdk1.8.0_221\bin\java.exe" ...
2020-05-19 16:03:32,387 [main] DEBUG [org.apache.ibatis.logging.LogFactory] - Logging initialized using 'class org.apache.ibatis
 .logging.log4j.Log4jImpl' adapter.
2020-05-19 16:03:32,491 [main] DEBUG [org.apache.ibatis.datasource.pooled.PooledDataSource] - PooledDataSource forcefully
 closed/removed all connections.
2020-05-19 16:03:32,491 [main] DEBUG [org.apache.ibatis.datasource.pooled.PooledDataSource] - PooledDataSource forcefully
 closed/removed all connections.
2020-05-19 16:03:32,492 [main] DEBUG [org.apache.ibatis.datasource.pooled.PooledDataSource] - PooledDataSource forcefully
 closed/removed all connections.
2020-05-19 16:03:32,492 [main] DEBUG [org.apache.ibatis.datasource.pooled.PooledDataSource] - PooledDataSource forcefully
 closed/removed all connections.
2020-05-19 16:03:32,878 [main] DEBUG [org.apache.ibatis.transaction.jdbc.JdbcTransaction] - Opening JDBC Connection
2020-05-19 16:03:33,564 [main] DEBUG [org.apache.ibatis.datasource.pooled.PooledDataSource] - Created connection 315932542.
2020-05-19 16:03:33,565 [main] DEBUG [org.apache.ibatis.transaction.jdbc.JdbcTransaction] - Setting autocommit to false on JDBC
 Connection [com.mysql.cj.jdbc.ConnectionImpl@12d4bf7e]
2020-05-19 16:03:33,579 [main] DEBUG [com.bailiban.mapper.UserMapper.findUserByName] - ==>  Preparing: select * from tab_user
 where username = ?
2020-05-19 16:03:33,745 [main] DEBUG [com.bailiban.mapper.UserMapper.findUserByName] - ==> Parameters: 百里半(String)
2020-05-19 16:03:33,869 [main] DEBUG [com.bailiban.mapper.UserMapper.findUserByName] - <==      Total: 1
User(id=45, username=百里半, birthday=Tue Jan 15 00:00:00 CST 2013, address=北京海淀区, sex=男)
2020-05-19 16:03:33,872 [main] DEBUG [org.apache.ibatis.transaction.jdbc.JdbcTransaction] - Resetting autocommit to true on JDBC
 Connection [com.mysql.cj.jdbc.ConnectionImpl@12d4bf7e]
2020-05-19 16:03:33,873 [main] DEBUG [org.apache.ibatis.transaction.jdbc.JdbcTransaction] - Closing JDBC Connection [com.mysql.cj
 .jdbc.ConnectionImpl@12d4bf7e]
2020-05-19 16:03:33,873 [main] DEBUG [org.apache.ibatis.datasource.pooled.PooledDataSource] - Returned connection 315932542 to
 pool.

Process finished with exit code 0
```

图 2-10

通过 log4j.properties 日志配置文件，控制台显示了 MyBatis 的整个组件以及 SQL、SQL 参数、返回结果等详细信息，开发者通过这些信息可以监控 MyBatis 的运行过程，定位问题的所在。

## ◆ 第二阶段◆

### 练习 2：使用注解实现映射器方式完成查询

**【问题描述】**

使用注解的方式根据用户的名称查询用户信息。

**【问题分析】**

根据文中案例，使用@Select 注解，传递一个 name 参数完成查询操作。

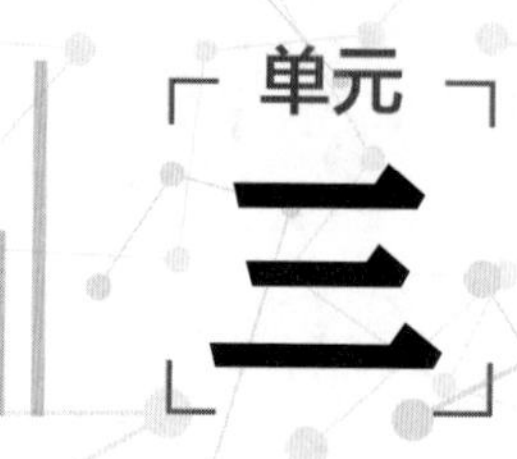


# MyBatis核心配置

## 课程目标

❖ 掌握 MyBatis 配置文件的编写

❖ 掌握 typeAliases 自定义别名的方法

❖ 掌握运行环境的配置

❖ 掌握第三方数据源的使用

**简介**

在上一单元中，我们已经用到了 MyBatis 的配置文件 mybatis-config.xml，其中包含了很多影响 MyBatis 行为的重要信息。因此熟悉配置文件中各个元素的功能十分重要。在 MyBatis 框架的核心配置文件中，<configuration>元素是配置文件的根元素，其他元素都要在<configuration>元素内配置。下面就来学习 MyBatis 核心配置文件。

## 3.1 MyBatis 配置文件结构

MyBatis 配置文件并不复杂，但是其存在一定的顺序结构，如果配置项顺序颠倒，MyBatis 启动阶段就会报错。下面的代码展示了 MyBatis 配置文件的结构。

```
<?xml version="1.0" encoding="UTF-8"?>
<!DOCTYPE configuration PUBLIC "-//mybatis.org//DTD Config 3.0//EN"
              "http://mybatis.org/dtd/mybatis-3-config.dtd">
<configuration><!--配置-->

    <properties resource="db.properties"/><!--属性-->
    <settings><!--设置-->
        <setting name="cacheEnabled" value="true"/>
        <setting name="defaultStatementTimeout" value="60"/>
    </settings>
   <typeAliases><!--类型别名-->
       <package name="com.bailiban.pojo"/>
   </typeAliases>
    <typeHandlers/><!--类型处理器-->
    <objectFactory  type=""/><!--对象工厂-->
    <plugins><!--插件-->
        <plugin interceptor=""></plugin>
    </plugins>
    <!-- 配置 MyBatis 运行环境 -->
    <environments default="development">
        <environment id="development"><!--环境变量-->
            <transactionManager type="JDBC"/><!--事务管理器-->
            <dataSource type="POOLED"><!--数据源-->
                <property name="driver" value="${jdbc.driver}"/>
                <property name="url" value="${jdbc.url}"/>
                <property name="username" value="${jdbc.username}"/>
                <property name="password" value="${jdbc.password}"/>
            </dataSource>
        </environment>
```

```
    </environments>
    <mappers>
        <!-- 映射文件 -->
        <mapper resource="com/bailiban/mapper/UserMapper.xml"/>
    </mappers>
</configuration>
```

这些配置项中，并不是所有的都是常用的，后面围绕常用的几个核心配置展开讲解。

# 3.2 properties 属性

properties 继承于 Hashtable，表示一个持久的属性集。属性列表中每个键及其对应值都是一个字符串。在 MyBatis 核心配置文件中，使用 property 子元素同样可以配置一些运行参数，这些参数既可以写在 XML 文件中，也可以写在 properties 文件中，然后在 XML 中引用。

通常在生产环境中数据库的密码等核心信息不会以明文的方式暴露，因此还会涉及 properties 文件中的数据加密解密。下面会逐一展开讨论。

## 3.2.1 property 子元素

<properties>元素下可以使用<property>子元素进行键值对的配置，代码如下所示。

```
<?xml version="1.0" encoding="UTF-8"?>
<!DOCTYPE configuration PUBLIC "-//mybatis.org//DTD Config 3.0//EN"
        "http://mybatis.org/dtd/mybatis-3-config.dtd">
<configuration><!--配置-->
    <!--属性-->
    <properties>
        <property name="jdbc.driver" value="com.mysql.cj.jdbc.Driver"/>
        <property name="jdbc.url"
            value="jdbc:mysql://localhost:3306/mybatis?serverTimezone
            =GMT%2b8&amp;useSSL=false&amp;useUnicode=true&amp;characterEncoding=utf-8"/>
        <property name="jdbc.username" value="root"/>
        <property name="jdbc.password" value="123456"/>
    </properties>
    <typeAliases><!--类型别名-->
        <package name="com.bailiban.pojo"/>
    </typeAliases>
    <!-- 配置 MyBatis 运行环境 -->
    <environments default="development">
        <environment id="development"><!--环境变量-->
```

```
            <transactionManager type="JDBC"/><!--事务管理器-->
            <dataSource type="POOLED"><!--数据源-->
                <property name="driver" value="${jdbc.driver}"/>
                <property name="url" value="${jdbc.url}"/>
                <property name="username" value="${jdbc.username}"/>
                <property name="password" value="${jdbc.password}"/>
            </dataSource>
        </environment>
    </environments>
    <mappers>
        <!-- 映射文件 -->
        <mapper resource="com/bailiban/mapper/UserMapper.xml"/>
    </mappers>
</configuration>
```

上面代码中在<properties>属性里面定义好了数据库连接信息，然后在<dataSource>下面使用<property>元素结合${}来引用上面定义好的属性值。这种做法看似重复，但是当多处需要用到定义好的 properties 属性时，直接引用即可。

但是这种写法并不推荐使用，因为当配置项比较多时，将定义属性和引用属性放在一个文件，就会显得混乱，并且也只有当前文件中才可以引用定义好的属性。此时就需要使用 properties 文件来解决。

## 3.2.2 使用 properties 文件

使用 properties 文件可将在 XML 中使用<properties>属性定义的内容抽离出来，写在 properties 文件中。在 src/main/resources 目录下新建 db.properties 文件，文件中代码如下所示：

```
jdbc.driver=com.mysql.cj.jdbc.Driver
jdbc.url=jdbc:mysql://localhost:3306/mybatis?serverTimezone=GMT%2b8&useSSL
        =false&useUnicode=true&characterEncoding=utf-8
jdbc.username=root
jdbc.password=123456
```

这样在 mybatis-config.xml 中就可以使用<properties>元素的 resource 属性来直接引入上面的 properties 文件。代码如下所示：

```
<properties resource="db.properties"/>
```

然后<dataSource>中关于数据库的配置，就和 3.2.1 小节中一样，使用${}的方式引用定义好的属性参数。后面只需要维护 db.properties 就可以维护我们的配置内容了。

### 3.2.3 properties 加密解密

在前面我们提到过在生产环境中，为了安全起见，数据库用户名和密码都会进行加密处理。因此配置在 properties 文件中的用户名和密码不能直接传给数据源，还需要进行解密操作。下面进行简单的模拟演示。db.properties 文件内容如下：

```
jdbc.driver=com.mysql.cj.jdbc.Driver
jdbc.url=jdbc:mysql://localhost:3306/mybatis?serverTimezone=GMT%2b8&useSSL
    =false&useUnicode=true&characterEncoding=utf-8
jdbc.username=cm9vdA==
jdbc.password=MTIzNDU2
```

此时看到的用户名和密码不再是简单的 root 和 123456。下面修改在单元二中创建的 SqlSessionFactoryWithXmlUtil 工具类。代码如下：

```
public class SqlSessionFactoryWithXmlUtil {

    /**
     * 获取当前对象用于加锁
     */
    private static final Class<SqlSessionFactoryWithXmlUtil> CLASS_LOCK =
            SqlSessionFactoryWithXmlUtil.class;
    /**
     * 首先创建静态成员变量 sqlSessionFactory，静态变量被所有的对象共享。
     */
    private static SqlSessionFactory sqlSessionFactory = null;

    /**
     * 构造私有函数
     */
    private SqlSessionFactoryWithXmlUtil() {
    }

    /**
     * 单例模式
     *
     * @return SqlSessionFactory
     */
    public static SqlSessionFactory getSqlSessionFactory() {
        synchronized (CLASS_LOCK) {
            if (sqlSessionFactory == null) {
                // 1.创建构建者对象 builder
                SqlSessionFactoryBuilder builder = new SqlSessionFactoryBuilder();
                // 2.读取 MyBatis 配置文件
                InputStream input = null;
                InputStream in;
```

```
                try {
                    in = Resources.getResourceAsStream("db.properties");
                    Properties properties = new Properties();
                    properties.load(in);
                    String username = properties.getProperty("jdbc.username");
                    String password = properties.getProperty("jdbc.password");
                    // 将加密的用户名和密码进行解密
                    byte[] decodeUsername = Base64.getDecoder().decode(username);
                    username = new String(decodeUsername, StandardCharsets.UTF_8);
                    byte[] decodePassword = Base64.getDecoder().decode(password);
                    password = new String(decodePassword, StandardCharsets.UTF_8);
                    // 重置 properties
                    properties.put("jdbc.username", username);
                    properties.put("jdbc.password", password);
                    input = Resources.getResourceAsStream("mybatis-config.xml");
                    // 3.使用构建者创建工厂对象 SqlSessionFactory
                    sqlSessionFactory = builder.build(input, properties);
                } catch (IOException e) {
                    e.printStackTrace();
                }

            }
            return sqlSessionFactory;
        }
    }

}
```

上面代码中通过 Resources 对象读取 db.properties 配置文件，然后拿到加密后的用户名和密码，并利用解密工具进行解密操作，然后重置 properties 文件。最后将 properties 传参给 SqlSessionFactoryBuilder 的 build 方法中。这样数据源中拿到原始的数据库用户名和密码，就可以连接上了。

## 3.3 settings 设置

settings 是 MyBatis 中极为重要的调整设置，它们会改变 MyBatis 的运行时行为。表 3-1 描述了各项设置的含义、默认值等。

表 3-1

| 设置名 | 描述 | 有效值 | 默认值 |
| --- | --- | --- | --- |
| cacheEnabled | 全局性地开启或关闭所有映射器配置文件中已配置的任何缓存 | true \| false | true |
| lazyLoadingEnabled | 延迟加载的全局开关。当开启时，所有关联对象都会延迟加载。特定关联关系中可通过设置 fetchType 属性来覆盖该项的开关状态 | true \| false | false |
| aggressiveLazyLoading | 开启时，任一方法的调用都会加载该对象的所有延迟加载属性。否则，每个延迟加载属性会按需加载(参考 lazyLoadTriggerMethods) | true \| false | false(在 3.4.1 及之前的版本中默认为 true) |
| multipleResultSetsEnabled | 是否允许单个语句返回多结果集(需要数据库驱动支持) | true \| false | true |
| useColumnLabel | 使用列标签代替列名。实际表现依赖于数据库驱动，具体可参考数据库驱动的相关文档，或通过对比测试来观察 | true \| false | true |
| useGeneratedKeys | 允许 JDBC 支持自动生成主键，需要数据库驱动支持。如果设置为 true，将强制使用自动生成主键。尽管一些数据库驱动不支持此特性，但仍可正常工作(如 Derby) | true \| false | false |
| autoMappingBehavior | 指定 MyBatis 应如何自动映射列到字段或属性。NONE 表示关闭自动映射；PARTIAL 只会自动映射没有定义嵌套结果映射的字段。FULL 会自动映射任何复杂的结果集(无论是否嵌套) | NONE、PARTIAL、FULL | PARTIAL |

(续表)

| 设置名 | 描述 | 有效值 | 默认值 |
| --- | --- | --- | --- |
| autoMappingUnknownColumnBehavior | 指定发现自动映射目标未知列(或未知属性类型)的行为。<br>NONE：不做任何反应<br>WARNING：输出警告日志('org.apache.ibatis. session. AutoMappingUnknownColumnBehavior'的日志等级必须设置为 WARN)<br>FAILING：映射失败(抛出 SqlSessionException) | NONE、WARNING、FAILING | NONE |
| defaultExecutorType | 配置默认的执行器。SIMPLE 就是普通的执行器；REUSE 执行器会重用预处理语句(PreparedStatement)；BATCH 执行器不仅重用语句还会执行批量更新 | SIMPLE REUSE BATCH | SIMPLE |
| defaultStatementTimeout | 设置超时时间，它决定数据库驱动等待数据库响应的秒数 | 任意正整数 | 未设置(null) |
| defaultFetchSize | 为驱动的结果集获取数量(fetchSize)设置一个建议值。此参数只可以在查询设置中被覆盖 | 任意正整数 | 未设置(null) |
| safeRowBoundsEnabled | 是否允许在嵌套语句中使用分页(RowBounds)。如果允许使用，则设置为 false | true \| false | false |
| safeResultHandlerEnabled | 是否允许在嵌套语句中使用结果处理器(ResultHandler)。如果允许使用则设置为 false | true \| false | true |
| mapUnderscoreToCamelCase | 是否开启驼峰命名自动映射，即从经典数据库列名 A_COLUMN 映射到经典 Java 属性名 aColumn | true \| false | false |

(续表)

| 设置名 | 描述 | 有效值 | 默认值 |
| --- | --- | --- | --- |
| localCacheScope | MyBatis 利用本地缓存机制(Local Cache)防止循环引用和加速重复的嵌套查询。默认值为 SESSION，会缓存一个会话中执行的所有查询。若设置值为 STATEMENT，本地缓存将仅用于执行语句，对相同 SqlSession 的不同查询将不会进行缓存 | SESSION \| STATEMENT | SESSION |
| jdbcTypeForNull | 当没有为参数指定特定的 JDBC 类型时，空值的默认 JDBC 类型。某些数据库驱动需要指定列的 JDBC 类型，多数情况直接用一般类型即可，比如 NULL、VARCHAR 或 OTHER | JdbcType 常量，常用值为 NULL、VARCHAR 或 OTHER | OTHER |
| lazyLoadTriggerMethods | 指定对象的哪些方法触发一次延迟加载 | 用逗号分隔的方法列表 | equals、clone、hashCode、toString |
| defaultResultSetType | 指定语句默认的滚动策略(新增于 3.5.2 版本) | FORWARD_ONLY \| SCROLL_SENSITIVE \| SCROLL_INSENSITIVE \| DEFAULT(等同于未设置) | 未设置(null) |

从表 3-1 可以看出，settings 的配置项相当多，但是真正用到的只有几个，比如关于缓存的 cacheEnabled，关于级联延迟加载的 lazyLoadingEnabled 和 aggressiveLazyLoading，关于驼峰命名自动映射的 mapUnderscoreToCamelCase，关于执行器 defaultExecutorType 等。下面给出一个配置完整的 settings 元素的示例：

```
<settings>
  <setting name="cacheEnabled" value="true"/>
  <setting name="lazyLoadingEnabled" value="true"/>
  <setting name="multipleResultSetsEnabled" value="true"/>
  <setting name="useColumnLabel" value="true"/>
  <setting name="useGeneratedKeys" value="false"/>
  <setting name="autoMappingBehavior" value="PARTIAL"/>
```

```
    <setting name="autoMappingUnknownColumnBehavior" value="WARNING"/>
    <setting name="defaultExecutorType" value="SIMPLE"/>
    <setting name="defaultStatementTimeout" value="25"/>
    <setting name="defaultFetchSize" value="100"/>
    <setting name="safeRowBoundsEnabled" value="false"/>
    <setting name="mapUnderscoreToCamelCase" value="false"/>
    <setting name="localCacheScope" value="SESSION"/>
    <setting name="jdbcTypeForNull" value="OTHER"/>
    <setting name="lazyLoadTriggerMethods" value="equals,clone,hashCode,toString"/>
</settings>
```

# 3.4 typeAliases 别名

类型别名可为 Java 类型设置一个缩写名字。它仅用于 XML 配置，意在降低冗余的全限定类名书写。MyBatis 已经为 Java 常见类型默认指定了别名，我们也可以自定义别名。

## 3.4.1 自定义别名

在开发中，使用别名可以提高开发效率，简化配置，例如下面的代码：

```
<typeAliases><!--类型别名-->
    <typeAlias type="com.bailiban.pojo.User" alias="user"/>
    <typeAlias type="com.bailiban.pojo.Role" alias="role"/>
    <typeAlias type="com.bailiban.pojo.Account" alias="account"/>
</typeAliases>
```

上面的代码展示了将 POJO 类的全限定名使用别名代替，当这样配置时，user 可以用在任何使用 com.bailiban.pojo.User 的地方。注意，在 MyBatis 中别名不区分大小写。

当然除了上面这种写法，还可以指定一个包名，MyBatis 会在包名下面搜索需要的 Java Bean，代码如下。

```
<typeAliases><!--类型别名-->
    <package name="com.bailiban.pojo"/>
</typeAliases>
```

每一个在包 com.bailiban.pojo 中的 Java Bean，在没有注解的情况下，会使用 Bean 的首字母小写的非限定类名来作为它的别名。比如 com.bailiban.pojo.Role 的别名为 role；若有注解，则别名为其注解值，代码如下：

```
@Alias("role")
public class Role{
```

```
        ...
    }
```

## 3.4.2 系统定义别名

MyBatis 已经为 Java 常见类型默认指定了别名，可以直接使用。因为有一些基本数据类型和包装类型的名称一样(例如基本数据类型 byte 和包装类型 java.lang.Byte)，所以在基本的数据类型前面加了下划线“_”来区分(byte 别名就是_byte，java.lang.Byte 别名就是 byte)。常见的类型对应别名如表 3-2 所示。

表 3-2

| 别名 | 映射的类型 |
| --- | --- |
| _byte | byte |
| _long | long |
| _short | short |
| _int | int |
| _integer | int |
| _double | double |
| _float | float |
| _boolean | boolean |
| string | String |
| byte | Byte |
| long | Long |
| short | Short |
| int | Integer |
| integer | Integer |
| double | Double |
| float | Float |
| boolean | Boolean |
| date | Date |
| decimal | BigDecimal |
| bigdecimal | BigDecimal |

(续表)

| 别名 | 映射的类型 |
|---|---|
| object | Object |
| map | Map |
| hashmap | HashMap |
| list | List |
| arraylist | ArrayList |
| collection | Collection |
| iterator | Iterator |

## 3.5 environments 运行环境

environments 的作用是配置运行环境，也就是数据库信息，可以配置多个，其有两个可配的子元素，分别是事务管理器 transactionManager 和数据源 dataSource。在实际开发中，我们多数会采用 Spring 对数据源和数据库的事务进行管理，在后面 Spring 的相关单元中会进行讲解。下面代码展示了运行环境的配置。

```
<!-- 配置 MyBatis 运行环境 -->
<environments default="development">
    <environment id="development"><!--环境变量-->
        <transactionManager type="JDBC"/><!--事务管理器-->
        <dataSource type="POOLED"><!--数据源-->
            <property name="driver" value="${jdbc.driver}"/>
            <property name="url" value="${jdbc.url}"/>
            <property name="username" value="${jdbc.username}"/>
            <property name="password" value="${jdbc.password}"/>
        </dataSource>
    </environment>
</environments>
```

### 3.5.1 设置 environment

environments 元素节点可以配置多个 environment，也就意味着 MyBatis 可以配置多个运行环境。因为在开发系统过程中，会分多套环境，如开发环境、测试环境、正式生产环境，它们所用的数据库不一样。所以此时可以设置多个 environment，通过 environment 的属性 id 来对应环境。当需要用那套环境时，就将 environments 的 default 属性的值配置为对应环境的 id 值。代码如下：

```
<!-- 配置 MyBatis 运行环境 -->
<environments default="development">
    <!--开发环境-->
    <environment id="development">
        <!--事务管理器-->
        <transactionManager type="JDBC"/>
        <!--数据源-->
        <dataSource type="POOLED">
            <property name="driver" value="${jdbc.driver}"/>
            <property name="url" value="${jdbc.url}"/>
            <property name="username" value="${jdbc.username}"/>
            <property name="password" value="${jdbc.password}"/>
        </dataSource>
    </environment>
    <!--测试环境-->
    <environment id="test">
        <!--事务管理器-->
        <transactionManager type="JDBC"/>
        <!--数据源-->
        <dataSource type="POOLED">
            <property name="driver" value="com.mysql.cj.jdbc.Driver"/>
            <property name="url"
value="jdbc:mysql://localhost:3306/mybatis_test?serverTimezone=GMT%2b8"/>
            <property name="username" value="root"/>
            <property name="password" value="123456"/>
        </dataSource>
    </environment>
</environments>
```

## 3.5.2 transactionManager 事务管理器

在 MyBatis 中有两种类型的事务管理器，也就是 type="JDBC"或 type="MANAGED"。它们都需要实现接口 Transaction，可以查看它的以下源代码：

```
public interface Transaction {

  /**
   * Retrieve inner database connection.
   * @return DataBase connection
   * @throws SQLException
   */
  Connection getConnection() throws SQLException;
```

```
    /**
     * Commit inner database connection.
     * @throws SQLException
     */
    void commit() throws SQLException;

    /**
     * Rollback inner database connection.
     * @throws SQLException
     */
    void rollback() throws SQLException;

    /**
     * Close inner database connection.
     * @throws SQLException
     */
    void close() throws SQLException;

    /**
     * Get transaction timeout if set.
     * @throws SQLException
     */
    Integer getTimeout() throws SQLException;

}
```

从提供的方法中可以知道，它主要完成提交、回滚、关闭数据库的事务。MyBatis 中为 Transaction 接口提供了两个实现类，分别是 JdbcTransaction 和 ManagedTransaction，并且分别对应着 JdbcTransactionFactory 和 ManagedTransactionFactory 两个工厂，这两个工厂实现了 TransactionFactory 这个接口，当在配置文件中通过 transactionManager 的 type 属性配置事务管理器类型的时候，MyBatis 就会自动从对应的工厂获取实例。下面介绍这两者的区别：

- JDBC 使用 JdbcTransactionFactory 工厂生成的 JdbcTransaction 对象实现，以 JDBC 的方式进行数据库的提交、回滚等操作。
- MANAGED 使用 ManagedTransactionFactory 工厂生成的 ManagedTransaction 对象实现，它的提交和回滚不需要任何操作，而是把事务交给容器进行处理，默认情况下会关闭连接，如果不希望默认关闭，只要将其中的 closeConnection 属性设置为 false 即可。

下面添加一个 User 来测试这两种事务配置的区别。代码如下所示：

```
@Test
public void insert() {
```

```
        SqlSession sqlSession = null;
        try {
            SqlSessionFactory sqlSessionFactory = SqlSessionFactoryWithXmlUtil.getSqlSessionFactory();
            // 生成 sqlSession
            sqlSession = sqlSessionFactory.openSession();
            // 使用 SqlSession 创建 mapper 接口的代理对象
            UserMapper mapper = sqlSession.getMapper(UserMapper.class);
            User user = new User();
            user.setAddress("武汉南湖");
            user.setBirthday(new Date());
            user.setSex("男");
            user.setUsername("陈建国");
            mapper.insert(user);
        } finally {
            if (sqlSession != null) {
                sqlSession.close();
            }
        }
    }
```

当 transactionManager 的事务配置为 JDBC 时，控制台显示的主要信息如图 3-1 所示。

```
2020-05-22 15:47:41,732 [main] DEBUG [org.apache.ibatis.transaction.jdbc.JdbcTransaction] - Opening JDBC Connection
2020-05-22 15:47:42,972 [main] DEBUG [org.apache.ibatis.datasource.pooled.PooledDataSource] - Created connection 315932542.
2020-05-22 15:47:42,973 [main] DEBUG [org.apache.ibatis.transaction.jdbc.JdbcTransaction] - Setting autocommit to false on JDBC
 Connection [com.mysql.cj.jdbc.ConnectionImpl@12d4bf7e]
2020-05-22 15:47:42,991 [main] DEBUG [com.bailiban.mapper.UserMapper.insert] - ==>  Preparing: insert into tab_user(username,
 birthday, sex, address) values(?,?,?,?)
2020-05-22 15:47:43,392 [main] DEBUG [com.bailiban.mapper.UserMapper.insert] - ==> Parameters: 陈建国(String), 2020-05-22
 15:47:41.625(Timestamp), 男(String), 武汉南湖(String)
2020-05-22 15:47:43,481 [main] DEBUG [com.bailiban.mapper.UserMapper.insert] - <==    Updates: 1
2020-05-22 15:47:43,483 [main] DEBUG [org.apache.ibatis.transaction.jdbc.JdbcTransaction] - Rolling back JDBC Connection
 [com.mysql.cj.jdbc.ConnectionImpl@12d4bf7e]
2020-05-22 15:47:43,496 [main] DEBUG [org.apache.ibatis.transaction.jdbc.JdbcTransaction] - Resetting autocommit to true on JDBC
 Connection [com.mysql.cj.jdbc.ConnectionImpl@12d4bf7e]
2020-05-22 15:47:43,497 [main] DEBUG [org.apache.ibatis.transaction.jdbc.JdbcTransaction] - Closing JDBC Connection [com.mysql.cj
 .jdbc.ConnectionImpl@12d4bf7e]
2020-05-22 15:47:43,497 [main] DEBUG [org.apache.ibatis.datasource.pooled.PooledDataSource] - Returned connection 315932542 to
 pool.
```

图 3-1

从图中标识部分可以看出，系统进行了回滚操作，查看数据库看不到刚才插入的数据，因为代码中未执行 sqlSession.commit()进行提交。

当 transactionManager 的事务配置为 MANAGED 时，控制台输出的主要信息如图 3-2 所示。

```
2020-05-22 15:53:35,766 [main] DEBUG [org.apache.ibatis.transaction.managed.ManagedTransaction] - Opening JDBC Connection
2020-05-22 15:53:37,432 [main] DEBUG [org.apache.ibatis.datasource.pooled.PooledDataSource] - Created connection 296347592.
2020-05-22 15:53:37,463 [main] DEBUG [com.bailiban.mapper.UserMapper.insert] - ==>  Preparing: insert into tab_user(username,
 birthday, sex, address) values(?,?,?,?)
2020-05-22 15:53:37,858 [main] DEBUG [com.bailiban.mapper.UserMapper.insert] - ==> Parameters: 陈建国(String), 2020-05-22
 15:53:35.595(Timestamp), 男(String), 武汉南湖(String)
2020-05-22 15:53:37,882 [main] DEBUG [com.bailiban.mapper.UserMapper.insert] - <==    Updates: 1
2020-05-22 15:53:37,883 [main] DEBUG [org.apache.ibatis.transaction.managed.ManagedTransaction] - Closing JDBC Connection
 [com.mysql.cj.jdbc.ConnectionImpl@11a9e7c8]
2020-05-22 15:53:37,884 [main] DEBUG [org.apache.ibatis.datasource.pooled.PooledDataSource] - Returned connection 296347592 to
 pool.
```

图 3-2

此时插入成功，查看数据库可以看到刚才插入的数据。

由此可以得出，当事务管理器使用 MANAGED 方式时，无须手动提交；当使用 JDBC 时，则需要进行 commit 操作。

### 3.5.3　dataSource 数据源

在MyBatis中，数据库通过PooledDataSourceFactory、UnpooledDataSourceFactory 和 JndiDataSourceFactory 三个工厂类来提供，前两者分别产生 PooledDataSource 和 UnpooledDataSource类对象，第三个则会根据JNDI的信息获取外部容器实现的数据库连接对象，但是不管怎样，它们最后都会生成一个实现了DataSource接口的数据库连接对象。

它们的配置信息如下：

```
<dataSource type="POOLED">
<dataSource type="UNPOOLED">
<dataSource type="JNDI">
```

这三种数据源的意义如下：

#### 1 UNPOOLED

采用非数据库池的管理方式，每次请求都会新建一个连接，所以性能不是很高，使用这种数据源的时候，可以配置以下属性：

- driver：数据库驱动名。
- url：数据库连接 URL。
- username：用户名。
- password：密码。
- defaultTransactionIsolationLevel：默认的事务隔离级别，如果要传递属性给驱动，则属性的前缀为 driver，例如：driver.encoding=UTF8。

#### 2. POOLED

采用连接池的概念将数据库连接对象 Connection 组织起来，可以在初始化时创建多个连接，使用时直接从连接池获取，避免了重复创建连接所需的初始化和认证时间，从而提升了效率，所以这种方式比较适合用在对性能要求高的应用中。除了 UNPOOLED 中的配置属性之外，还有下面几个针对池子的配置：

- poolMaximumActiveConnections：任意时间都会存在的连接数，默认值为 10。
- poolMaxmumIdleConnections：可以空闲存在的连接数。

- poolMaxmumCheckoutTime：在被强制返回之前，检查出存在空闲连接的等待时间。即如果有 20 个连接，只有一个空闲，在这个空闲连接被找到之前的等待时间就用这个属性配置。
- poolTimeToWait：等待一个数据库连接成功所需的时间，如果超出这个时间则尝试重新连接。
- poolPingQuery：发送到数据库的侦测查询，用来检验连接是否处在正常工作秩序中并准备接受请求。默认是 NO PING QUERY SET，这会导致多数数据库驱动失败时带有一个恰当的错误消息。
- poolPingEnabled：是否启用侦测查询。若开启，也必须使用一个可执行的 SQL 语句设置 poolPingQuery 属性(最好是一个非常快的 SQL)，默认值为 false。
- poolPingConnectionsNotUsedFor：配置 poolPingQuery 的使用频度。这可以被设置成匹配具体的数据库连接超时时间，来避免不必要的侦测，默认值为 0(即所有连接每一时刻都被侦测，当然仅当 poolPingEnabled 为 true 时适用)。

3. JNDI

数据源 JNDI 的实现是为了能在如 EJB 或应用服务器这类容器中使用，容器可以集中或在外部配置数据源，然后放置一个 JNDI 上下文的引用。这种数据源只需配置两个属性：

- initial_context：用来在 InitialContext 中寻找上下文。可选，如果忽略，data_source 属性将会直接从 InitialContext 中寻找。
- data_source：引用数据源实例位置上下文的路径。当提供 initial_context 配置时，data_source 会在其返回的上下文查找，否则直接从 InitialContext 中查找。

和其他数据源配置类似，可以通过添加前缀 env.直接把属性传递给初始上下文。比如：env.encoding=UTF8，就会在初始上下文(InitialContext)实例化时往它的构造方法传递值为 UTF8 的 encoding 属性。

除了上述三种数据源之外，MyBatis 还提供第三方数据源，如阿里巴巴的 Druid，但是需要我们自定义数据源工厂并进行配置。

首先使用 Druid，需要在 pom.xml 文件中加入其依赖坐标，如下：

```
<!-- https://mvnrepository.com/artifact/com.alibaba/druid -->
<dependency>
    <groupId>com.alibaba</groupId>
    <artifactId>druid</artifactId>
    <version>1.1.20</version>
</dependency>
```

在 com.bailiban.datasource 包下新建一个 DruidDataSourceFactory 类，继承 org.apache.ibatis.datasource.unpooled.UnpooledDataSourceFactory。代码如下所示：

```
public class DruidDataSourceFactory extends UnpooledDataSourceFactory {
    public DruidDataSourceFactory() {
        this.dataSource = new DruidDataSource();
    }
}
```

上面的代码十分简单，是一个在构造方法中将父类中的 dataSource 属性赋值为 DruidDataSource 的一个实例。

然后修改 mybatis-config.xml 中的数据源信息，代码如下：

```
<!--数据源-->
<dataSource type="com.bailiban.datasource.DruidDataSourceFactory">
    <property name="url" value="${jdbc.url}" />
    <property name="username" value="${jdbc.username}" />
    <property name="password" value="${jdbc.password}" />
    <!-- 配置初始化大小、最小、最大 -->
    <property name="initialSize" value="1" />
    <property name="minIdle" value="1" />
    <property name="maxActive" value="20" />
</dataSource>
```

从上面的代码可以发现，只需将 <dataSource> 元素的 type 属性设置为上面编写的 DruidDataSourceFactory 类型即可。这样就完成了使用第三方数据源的配置。

## 单元小结

- MyBatis 核心配置文件结构存在一定的顺序。
- <property>元素的 value 属性可以结合${}来引用配置中定义好的 properties 属性值。
- MyBatis 已经为 Java 常见类型默认指定了别名。
- MyBatis 可以配置多套运行环境。
- MyBatis 中，数据库是通过 PooledDataSourceFactory、UnpooledDataSourceFactory 和 JndiDataSourceFactory 三个工厂类来提供。

## 单元自测

1. MyBatis有两种事务管理器类型，分别是(　　)和(　　)。

A. JDBC　　B. MANAGED

C. POOLED　　D. JNDI

2. 以下属于MyBatis内置类型别名的有(　　)。

A. _int　　B. Integer

C. int　　D. String

3. MyBatis框架自身提供了三种数据源，但(　　)不属于。

A. POOLED　　B. UNPOOLED

C. JNDI　　D. Druid

## 上机实战

### 上机目标

- 掌握第三方数据源的使用。

### 上机练习

◆ 第一阶段◆

#### 练习1：阿里巴巴旗下的Druid数据源的使用

**【问题描述】**

使用第三方数据源Druid完成数据的查询。

**【问题分析】**

将dataSource配置成第三方数据源。

**【参考步骤】**

(1) 在com.bailiban.datasource下创建DruidDataSourceFactory并继承MyBatis的UnpooledDataSourceFactory类，代码如下所示。

```
public class DruidDataSourceFactory extends UnpooledDataSourceFactory {
    public DruidDataSourceFactory() {
        this.dataSource = new DruidDataSource();
    }
}
```

(2) 修改 mybatis-config.xml 配置，代码如下。

```
<?xml version="1.0" encoding="UTF-8"?>
<!DOCTYPE configuration PUBLIC "-//mybatis.org//DTD Config 3.0//EN"
"http://mybatis.org/dtd/mybatis-3-config.dtd">
<configuration><!--配置-->
    <!--属性-->
    <properties resource="db.properties"/>
    <typeAliases><!--类型别名-->
        <package name="com.bailiban.pojo"/>
    </typeAliases>
    <!-- 配置 MyBatis 运行环境 -->
    <environments default="development">
        <!--开发环境-->
        <environment id="development">
            <!--事务管理器-->
            <transactionManager type="JDBC"/>
            <!--数据源-->
            <dataSource type="com.bailiban.datasource.DruidDataSourceFactory">
                <property name="url" value="${jdbc.url}" />
                <property name="username" value="${jdbc.username}" />
                <property name="password" value="${jdbc.password}" />
                <!-- 配置初始化大小、最小、最大 -->
                <property name="initialSize" value="1" />
                <property name="minIdle" value="1" />
                <property name="maxActive" value="20" />
            </dataSource>
        </environment>
    </environments>
    <mappers>
        <!-- 映射文件 -->
        <mapper resource="com/bailiban/mapper/UserMapper.xml"/>
    </mappers>
</configuration>
```

(3) 在测试类中进行测试，代码如下。

```
@Test
public void findUserByName() {
    SqlSession sqlSession = null;
    try {
        SqlSessionFactory sqlSessionFactory = SqlSessionFactoryWithXmlUtil.getSqlSessionFactory();
        // 生成 sqlSession
        sqlSession = sqlSessionFactory.openSession();
        // 使用 SqlSession 创建 mapper 接口的代理对象
        UserMapper mapper = sqlSession.getMapper(UserMapper.class);
        User user = mapper.findUserByName("百里半");
        System.out.println(user);
```

```
        } finally {
            if (sqlSession != null) {
                sqlSession.close();
            }
        }
    }
```

(4) 运行结果如图 3-3 所示。

```
2020-05-22 17:20:18,608 [main] DEBUG [org.apache.ibatis.transaction.jdbc.JdbcTransaction] - Opening JDBC Connection
2020-05-22 17:20:19,665 [main] INFO  [com.alibaba.druid.pool.DruidDataSource] - {dataSource-1} inited
2020-05-22 17:20:19,694 [main] DEBUG [org.apache.ibatis.transaction.jdbc.JdbcTransaction] - Setting autocommit to false on JDBC
 Connection [com.mysql.cj.jdbc.ConnectionImpl@61862a7f]
2020-05-22 17:20:19,728 [main] DEBUG [com.bailiban.mapper.UserMapper.findUserByName] - ==>  Preparing: select * from tab_user
 where username = ?
2020-05-22 17:20:20,030 [main] DEBUG [com.bailiban.mapper.UserMapper.findUserByName] - ==> Parameters: 百里半(String)
2020-05-22 17:20:20,167 [main] DEBUG [com.bailiban.mapper.UserMapper.findUserByName] - <==      Total: 1
User(id=45, username=百里半, birthday=Tue Jan 15 00:00:00 CST 2013, address=北京海淀区, sex=男)
2020-05-22 17:20:20,170 [main] DEBUG [org.apache.ibatis.transaction.jdbc.JdbcTransaction] - Resetting autocommit to true on JDBC
 Connection [com.mysql.cj.jdbc.ConnectionImpl@61862a7f]
2020-05-22 17:20:20,171 [main] DEBUG [org.apache.ibatis.transaction.jdbc.JdbcTransaction] - Closing JDBC Connection [com.mysql.cj
 .jdbc.ConnectionImpl@61862a7f]
```

图 3-3

## ◆第二阶段◆

### 练习 2：自定义别名的使用

**【问题描述】**

使用自定义别名完成 UserMapper.xml 中查询用户的编写。

**【问题分析】**

考察别名的使用，关键在于别名的定义，可以对实体类一个个配置别名，也可以使用 package 定义指定包下的 pojo 统一定义别名。

**【参考步骤】**

(1) 在 mybatis-config.xml 中使用 typeAliases 定义别名。

(2) 在映射文件 UserMapper.xml 中使用自定义的别名。

# 映 射 器

- 了解 ASP.NET MVC 开发模式
- 掌握 select、insert、delete 和 update 元素的使用方法
- 掌握参数传递的方法和指定返回参数类型
- 掌握级联操作

**简介**

MyBatis 的真正强大在于它的语句映射，在上个单元我们也了解到映射器既可以使用 XML 方式实现，也可以使用注解方式实现，但是对于复杂的 SQL 语句和级联操作，XML 方式更加灵活简单易于配置。因此本书以 XML 方式为主。使用 XML 映射器方式与具有相同功能的 JDBC 代码进行对比，能省掉将近 95% 的代码。MyBatis 致力于减少使用成本，让用户能更专注于 SQL 代码。

在 MyBatis 应用开发中，映射器的编写是最重要的一环，目前对于一些简单的 SQL 编写可以使用插件完成，例如 Idea 中的 Easy Code 插件可以自动完成映射器 Mapper 的编写。关于插件的用法，将在本书最后一个单元中进行展示。

下面展示了映射器的配置元素，如表 4-1 所示。

表 4-1

| 元素名称 | 描述 | 备注 |
| --- | --- | --- |
| cache | 该命名空间的缓存配置 | |
| cache-ref | 引用其他命名空间的缓存配置 | |
| resultMap | 描述如何从数据库结果集中加载对象，是最复杂也是最强大的元素 | 提供映射规则 |
| parameterMap | 老式风格的参数映射 | 此元素已被废弃，并可能在将来被移除 |
| sql | 可被其他语句引用的可重用语句块 | 一次定义，可以在多个 SQL 语句中使用 |
| insert | 映射插入语句 | 执行后返回一个整数，代表插入的条数 |
| update | 映射更新语句 | 执行后返回一个整数，代表更新的条数 |
| delete | 映射删除语句 | 执行后返回一个整数，代表删除的条数 |
| select | 映射查询语句 | 可以自定义参数，返回结果集等 |

## 4.1 引入映射器的方法

在 MyBatis 的配置文件中引入映射器，也就是告诉 MyBatis 到哪里去找映射文件。引入映射器的方法有下面 4 种。

(1) 使用相对于类路径的资源引用。

```
<mappers>
```

```
        <!--使用相对于类路径的资源引用-->
        <mapper resource="com/bailiban/mapper/UserMapper.xml"/>
</mappers>
```

(2) 使用完全限定资源定位符(URL)。

```
<mappers>
        <!--使用完全限定资源定位符(URL) -->
        <mapper url="file:///D:/myIdeaCode/MyBatis-demo2/src/main/resources/com/bailiban/mapper/
                        UserMapper.xml"/>
</mappers>
```

(3) 使用映射器接口实现类的完全限定类名。

```
<mappers>
        <!--使用映射器接口实现类的完全限定类名-->
        <mapper class="com.bailiban.mapper.UserMapper"/>
</mappers>
```

(4) 用包引入映射器。

```
<mappers>
        <!--将包内的映射器接口实现全部注册为映射器-->
  <package name="com.bailiban.mapper"/>
</mappers>
```

这里推荐使用第一和第四种方式引入映射器。

## 4.2 select 元素

查询语句是 MyBatis 中最常用的元素之一，仅仅把数据存到数据库中价值并不大，还要能重新取出来才有用，多数应用也都是查询比修改要频繁。MyBatis 的基本原则之一是：在每个插入、更新或删除操作之间，通常会执行多个查询操作。因此，MyBatis 在查询和结果映射做了相当多的改进。一个简单查询的 select 元素是非常简单的。

我们先看看 select 元素的属性，如表 4-2 所示。

表 4-2

| 元素 | 说明 |
|---|---|
| id | 在命名空间中唯一的标识符，可以被用来引用这条语句 |
| parameterType | 将会传入这条语句的参数的类全限定名或别名。这个属性是可选的，因为 MyBatis 可以通过类型处理器(TypeHandler)推断出具体传入语句的参数，默认值为未设置(unset) |

(续表)

| 元素 | 说明 |
| --- | --- |
| parameterMap | 用于引用外部parameterMap的属性，目前已被废弃。推荐使用行内参数映射和parameterType属性 |
| resultType | 期望从这条语句中返回结果的类全限定名或别名。注意，如果返回的是集合，那应该设置为集合包含的类型，而不是集合本身的类型。resultType 和 resultMap 之间只能同时使用一个 |
| resultMap | 对外部 resultMap 的命名引用。结果映射是 MyBatis 最强大的特性，若对其理解透彻，许多复杂的映射问题都能迎刃而解。resultType 和 resultMap 之间只能同时使用一个 |
| flushCache | 将其设置为 true 后，只要语句被调用，都会导致本地缓存和二级缓存被清空，默认值为 false |
| useCache | 将其设置为 true 后，将会导致本条语句的结果被二级缓存缓存起来，默认情况下，设置 select 元素为 true |
| timeout | 这个设置是在抛出异常之前，驱动程序等待数据库返回请求结果的秒数。默认值为未设置(unset)(依赖数据库驱动) |
| fetchSize | 这是一个给驱动的建议值，尝试让驱动程序每次批量返回的结果行数等于这个设置值。默认值为未设置(unset)(依赖驱动) |
| statementType | 可选 STATEMENT、PREPARED 或 CALLABLE，这会让 MyBatis 分别使用 Statement、PreparedStatement 或 CallableStatement，默认值为 PREPARED |
| resultSetType | FORWARD_ONLY、SCROLL_SENSITIVE、SCROLL_INSENSITIVE 或 DEFAULT(等价于 unset) 中的一个，默认值为 unset (依赖数据库驱动) |
| databaseId | 如果配置了数据库厂商标识(databaseId Provider)，MyBatis 会加载所有不带 databaseId 或匹配当前 databaseId 的语句；如果带和不带的语句都有，则不带的会被忽略 |
| resultOrdered | 这个设置仅针对嵌套结果 select 语句：如果为 true，将会假设包含了嵌套结果集或是分组，当返回一个主结果行时，就不会产生对前面结果集的引用。这就使得在获取嵌套结果集的时候不至于内存不够用。默认值为 false |
| resultSets | 这个设置仅适用于多结果集的情况。它将列出语句执行后返回的结果集并赋予每个结果集一个名称，多个名称之间以逗号分隔 |

### 4.2.1 简单的 select 元素的应用

在前面我们已经接触过 select 元素，下面再演示一个简单的例子：查询用户表中与指定姓氏相同的人数。代码如下所示。

```
<select id="selectCountByFirstName" parameterType="string" resultType="int">
    select count(*)  from tab_user where username like concat (#{name},'%')
```

```
</select>
```

代码比较简单，对于其中的一些元素的含义解释如下：

- id：命名空间中的唯一标识符，可用来代表这条语句。
- parameterType：SQL 接收的参数类型，可以使用类全限定名或别名。参数可选，因此也可以不写，为了代码的可读性，建议写上。
- resultType：用于设置返回值的类型和映射关系，和 parameterType 一样，可以使用类全限定名或别名，在这里参数为必选。
- #{name}：被传递进去的参数。

写好 SQL 后，还需要有一个接口方法与之匹配。代码如下：

```
/**
 * 查询同一姓氏的用户
 * @param name 姓
 * @return count
 */
int selectCountByFirstName(@Param("name") String name);
```

这样就完成了一个简单的 select 元素的应用。

## 4.2.2　自动映射和驼峰映射

MyBatis 提供了自动映射功能，且在默认情况下是开启的。在 setting 中有两个可以配置的选项 autoMappingBehavior 和 mapUnderscoreToCamelCase，它们分别控制自动映射和驼峰式映射。在这里推荐使用自动映射，因为可以通过 SQL 的别名来处理一些细节，比较灵活，而驼峰映射则要求比较严苛，下面分别来进行讲解。

配置自动映射的 autoMappingBehavior 选项的含义如下：

- NONE，不进行自动映射。
- PARTIAL，默认值，只对没有嵌套的结果集进行自动映射。
- FULL，对所有的结果集进行自动映射，包括嵌套结果集。

在默认情况下，使用默认的 PARTIAL 级别就可以了。下面使用实体类 Role 来做演示。

在 com.bailiban.pojo 中创建 Role 实体类，代码如下：

```
@Data
public class Role {
    /**
     * id
     */
    private Integer id;
    /**
```

```
        * 角色名称
        */
       private String name;
       /**
        * 角色描述
        */
       private String description;

   }
```

对应的表结构如下：

```
CREATE TABLE `tab_role` (
  `ID` int(11) NOT NULL AUTO_INCREMENT COMMENT '编号',
  `ROLE_NAME` varchar(30) DEFAULT NULL COMMENT '角色名称',
  `ROLE_DESP` varchar(60) DEFAULT NULL COMMENT '角色描述',
  PRIMARY KEY (`ID`)
) ENGINE=InnoDB AUTO_INCREMENT=11 DEFAULT CHARSET=utf8;
```

这个时候可以看到我们的实体类的属性名和表中的列名不一致，这个时候自动映射并不能将查询结果映射到 Role 实体类上。此时可以通过 SQL 别名的方式使其一致，代码如下：

```
<select id="findRoleById" parameterType="integer" resultType="role">
    select id, ROLE_NAME name, ROLE_DESP description
    from tab_role
    where id = #{id}
</select>
```

上面代码中我们将 ROLE_NAME 取了别名 name，ROLE_DESP 取别名为 description，这样就和 Role 实体类上的属性名称保持一致了。这样 MyBatis 就可以完成自动映射。

在了解了自动映射后，下面介绍使用驼峰映射的做法。

所谓驼峰映射，指的就是表字段和 POJO 的属性名要按照驼峰命名法严格对应，就能完成自动映射。例如，tab_role 表字段 ROLE_NAME(不区分大小写)，则 POJO 属性名为 roleName；字段 ROLE_DESP，则 POJO 属性为 roleDesp。

要使用驼峰映射，需要在 MyBatis-config.xml 中将 mapUnderscoreToCamelCase 设置为 true。代码如下所示：

```
<settings>
    <setting name="mapUnderscoreToCamelCase" value="true"/>
</settings>
```

综上所述，我们会发现使用驼峰映射时，数据字段和 POJO 的属性名严格对应，这样降低了灵活性，因此这也是我们不推荐的原因。

## 4.2.3 传递多个参数

在前面的例子中，SQL 传递的都是一个参数，那对于多个参数该怎么做呢？这里有几种不同的解决方法，下面我们以 tab_user 表来演示：查询指定城市指定姓氏相同的用户信息。

### 1. 使用 map 接口传递参数

在 MyBatis 中允许使用 map 接口通过键值对传递多个参数，此时在接口中定义的方法代码如下：

```
/**
 * 查询指定地区同一姓氏的用户
 * @param map 传多个参数
 * @return users
 */
List<User> findUserByFirstNameAndAddress(Map<String,Object> map);
```

此时传递给映射器的是一个 map 对象，对应的 SQL 语句如下：

```
<select id="findUserByFirstNameAndAddress" resultType="User" parameterType="map">
    select * from tab_user where username like concat (#{name},'%') and address like concat
    ('%',#{address},'%')
</select>
```

上面代码中的参数 name、address 是 map 集合中的键，测试代码如下：

```
@Test
public void findUserByFirstNameAndAddress() {
    SqlSession sqlSession = null;
    try {
        SqlSessionFactory sqlSessionFactory = SqlSessionFactoryWithXmlUtil.getSqlSessionFactory();
        // 生成 sqlSession
        sqlSession = sqlSessionFactory.openSession();
        // 使用 SqlSession 创建 mapper 接口的代理对象
        UserMapper mapper = sqlSession.getMapper(UserMapper.class);
        Map<String, Object> map = new HashMap<>();
        map.putIfAbsent("name", "小");
        map.putIfAbsent("address", "武汉");
        List<User> users = mapper.findUserByFirstNameAndAddress(map);
        for (User user : users) {
            System.out.println(user);
        }
    } finally {
        if (sqlSession != null) {
            sqlSession.close();
        }
```

```
    }
}
```

使用 map 基本可以解决所有的参数问题，但是我们并不推荐，因为它的可读性比较差，且从接口无法直接看出map 中传递的是什么。接下来介绍如何使用注解方式传递多个参数。

### 2. 使用注解传递多个参数

使用注解方式接口方法的代码如下：

```
/**
 * 查询指定地区同一姓氏的用户
 * @param name 姓氏
 * @param address 地址
 * @return users
 */
List<User> findUserByFirstNameAndAddress(@Param("firstName") String name, @Param("address")
    String address);
```

代码中的@Param(org.apache.ibatis.annotations.Param)注解，就可以将 name、address 参数传给 SQL，这里注意，@Param 注解中的 name 名称并不要求和接口参数名一样，但是要和 SQL 中的参数名一致。SQL 代码如下：

```
<select id="findUserByFirstNameAndAddress" resultType="User" >
    select * from tab_user where username like concat (#{firstName},'%') and address like concat
        ('%',#{address},'%')
</select>
```

对于使用注解方式，此时就不用给出 parameterType 属性。当然我们前面已经提到过 parameterType 属性是可选的，在上一节使用 map 方式中的 parameterType 属性也是可以不给出的。

### 3. 通过 Java Bean 传递多个参数

下面直接通过传递一个 User 作为参数，接口方法代码如下：

```
/**
 * 查询指定地区同一姓氏的用户
 *
 * @param user 传递 pojo 参数
 * @return users
 */
List<User> findUserByFirstNameAndAddress(User user);
```

映射文件中的 SQL 代码如下：

```
<select id="findUserByFirstNameAndAddress" resultType="User" parameterType="User">
    select * from tab_user where username like concat (#{username},'%') and address like concat
```

```
('%',#{address},'%')
    </select>
```

SQL 中的 username、address 和 User 的属性名要一致。

上面列举的三种方式也可以混合使用，这里就不作展示了。

## 4.2.4 使用 resultMap 映射结果集

在 4.2.2 节中我们讲到使用自动映射和驼峰映射来映射结果集，但是对于一些复杂的 SQL，比如涉及级联时就无法满足开发的要求。为了支持复杂的映射，select 元素提供了 resultMap 属性。下面的代码展示了 resultMap 用法，代码如下：

```
<mapper namespace="com.bailiban.mapper.RoleMapper">
    <resultMap id="roleMap" type="Role">
        <id property="id" column="id"/>
        <result property="name" column="ROLE_NAME"/>
        <result property="description" column="ROLE_DESP"/>
    </resultMap>
    <select id="findRoleById" parameterType="integer" resultMap="roleMap">
        select id, ROLE_NAME, ROLE_DESP
        from tab_role
        where id = #{id}
    </select>
</mapper>
```

对上面的代码阐述如下：

- resultMap 元素定义了一个 roleMap，它的属性 id 代表它的标识，type 代表使用那个类作为其映射的类，可以是别名也可以是全限定名。
- 它的子元素 id 代表 resultMap 的主键，而 result 代表其属性，property 代表 POJO 的属性名称，而 column 代表 SQL 的列名。这样就可以把 POJO 的属性和 SQL 的列名做对应，形成映射关系。

resultMap 元素十分强大，关于其更多更复杂的用法我们会单独放到后面的 resultMap 元素中进行讲解。

## 4.2.5 分页参数 RowBounds

分页是我们开发中都会遇到的，MySql 中使用 limit 语句来分页，但 limit 并不是标准 SQL 中的，如果使用其它数据库，则需要更换成其他语句。MyBatis 提供了 RowBounds 类，用于实现分页查询。下面我们看一下 RowBounds 源码，代码如下所示：

```
package org.apache.ibatis.session;
```

```
/**
 * @author Clinton Begin
 */
public class RowBounds {

  public static final int NO_ROW_OFFSET = 0;
  public static final int NO_ROW_LIMIT = Integer.MAX_VALUE;
  public static final RowBounds DEFAULT = new RowBounds();

  private final int offset;
  private final int limit;

  public RowBounds() {
    this.offset = NO_ROW_OFFSET;
    this.limit = NO_ROW_LIMIT;
  }

  public RowBounds(int offset, int limit) {
    this.offset = offset;
    this.limit = limit;
  }

  public int getOffset() {
    return offset;
  }

  public int getLimit() {
    return limit;
  }

}
```

offset 属性是偏移量，即从第几行开始读取记录。limit 是限制条数。逻辑分页的实现原理可以查看 MyBatis 中的 DefaultResultSetHandler 类，截取部分源码如下：

```
private void handleRowValuesForSimpleResultMap(ResultSetWrapper rsw, ResultMap resultMap,
      ResultHandler<?> resultHandler, RowBounds rowBounds, ResultMapping parentMapping)
      throws SQLException {
    DefaultResultContext<Object> resultContext = new DefaultResultContext<Object>();
//跳过 RowBounds 设置的 offset 值
    skipRows(rsw.getResultSet(), rowBounds);
//判断数据是否小于 limit，如果小于 limit 的话就不断的循环取值
    while (shouldProcessMoreRows(resultContext, rowBounds) && rsw.getResultSet().next()) {
      ResultMap discriminatedResultMap = resolveDiscriminatedResultMap(rsw.getResultSet(),
            resultMap, null);
      Object rowValue = getRowValue(rsw, discriminatedResultMap);
```

```
            storeObject(resultHandler, resultContext, rowValue, parentMapping, rsw.getResultSet());
        }
    }
private boolean shouldProcessMoreRows(ResultContext<?> context, RowBounds rowBounds) throws
        SQLException {
//判断数据是否小于 limit，若小于则返回 true
      return !context.isStopped() && context.getResultCount() < rowBounds.getLimit();
    }
    //跳过不需要的行，即 rowbounds 设置的 limit 和 offset
    private void skipRows(ResultSet rs, RowBounds rowBounds) throws SQLException {
      if (rs.getType() != ResultSet.TYPE_FORWARD_ONLY) {
        if (rowBounds.getOffset() != RowBounds.NO_ROW_OFFSET) {
          rs.absolute(rowBounds.getOffset());
        }
      } else {
    //跳过 RowBounds 中设置的 offset 条数据
        for (int i = 0; i < rowBounds.getOffset(); i++) {
          rs.next();
        }
      }
    }
```

从中可以看出 RowBounds 在处理分页时，只是简单地把 offset 之前的数据都 skip 掉，超过 limit 之后的数据不取出。跳过 offset 之前的数据是由方法 skipRows 处理，数据是否超过了 limit，则是由 shouldProcessMoreRows 方法进行判断。简言之，就是先把数据全部查询到 ResultSet，然后从 ResultSet 中取出 offset 和 limit 之间的数据，这样就实现了分页查询。

下面以分页查询 tab_role 表来举例，接口方法代码如下所示：

```
/**
 * 分页查询
 * @param rowBounds 分页信息对象
 * @return roles
 */
List<Role> selectAllByRowBounds(RowBounds rowBounds);
```

映射文件 RoleMapper.xml 中的 SQL 代码如下：

```
<select id="selectAllByRowBounds" resultMap="roleMap">
    select *
    from tab_role
</select>
```

在映射文件中，没有任何关于 RowBounds 参数的信息，它是 MyBatis 的一个附加参数，MyBatis 会自动识别它，然后据此进行分页。测试代码如下所示：

```
@Test
public void selectAllByRowBounds() {
    SqlSession session = null;
    try {
        SqlSessionFactory sqlSessionFactory = SqlSessionFactoryWithXmlUtil.getSqlSessionFactory();
        session = sqlSessionFactory.openSession();
        // 使用 SqlSession 创建 mapper 接口的代理对象
        RoleMapper mapper = session.getMapper(RoleMapper.class);
        // offset 起始行，limit 指当前页显示多少条数据
        RowBounds rowBounds = new RowBounds(2, 5);
        List<Role> roles = mapper.selectAllByRowBounds(rowBounds);
        for (Role role : roles) {
            System.out.println(role);
        }
    } finally {
        if (session != null) {
            session.close();
        }
    }
}
```

上面代码标识的是从 tab_role 表中查询 5 条数据，并且是从第 3 条数据开始。使用 RowBounds 进行分页只适合小数据量的查询，因为它是查询出所有，然后从结果中按要求截取指定条目的数据进行返回。因此对于大量的数据查询，性能并不佳，此时可以通过分页插件去处理，在本书的最后一个单元中会详细讲解 PageHelper 分页插件的使用。

## 4.3 insert 元素

<insert>元素用于映射插入语句，MyBatis 执行完一条插入语句后将返回一个整数，表示其影响的行数。它的属性与<select>元素的属性大部分相同，如表 4-3 所示。

表 4-3

| 元素 | 说明 |
| --- | --- |
| id | 在命名空间中唯一的标识符，可以被用来引用这条语句 |
| parameterType | 将会传入这条语句的参数的类全限定名或别名。这个属性是可选的，因为 MyBatis 可以通过类型处理器(TypeHandler)推断出具体传入语句的参数，默认值为未设置(unset) |
| parameterMap | 用于引用外部 parameterMap 的属性，目前已被废弃。可使用行内参数映射和 parameterType 属性 |

(续表)

| 元素 | 说明 |
| --- | --- |
| flushCache | 将其设置为 true 后，只要语句被调用，都会导致本地缓存和二级缓存被清空，默认值(对 insert、update 和 delete 语句)为 true |
| timeout | 这个设置是在抛出异常之前，驱动程序等待数据库返回请求结果的秒数。默认值为未设置(unset)(依赖数据库驱动) |
| statementType | 可选 STATEMENT、PREPARED 或 CALLABLE。这会让 MyBatis 分别使用 Statement、PreparedStatement 或 CallableStatement，默认值为 PREPARED |
| useGeneratedKeys | (仅适用于 insert 和 update)这会令 MyBatis 使用 JDBC 的 getGeneratedKeys 方法来取出由数据库内部生成的主键(比如：像 MySQL 和 SQL Server 这样的关系型数据库管理系统的自动递增字段)，默认值为 false |
| keyProperty | (仅适用于 insert 和 update)指定能够唯一识别对象的属性，MyBatis 会使用 getGeneratedKeys 的返回值或 insert 语句的 selectKey 子元素设置它的值，默认值为未设置(unset)。如果生成列不止一个，可以用逗号分隔多个属性名称 |
| keyColumn | (仅适用于 insert 和 update)设置生成键值在表中的列名，在某些数据库(像 PostgreSQL)中，当主键列不是表中的第一列的时候，是必须设置的。如果生成列不止一个，可以用逗号分隔多个属性名称 |
| databaseId | 如果配置了数据库厂商标识(databaseId Provider)，MyBatis 会加载所有不带 databaseId 或匹配当前 databaseId 的语句；如果带和不带的语句都有，则不带的会被忽略 |

## 4.3.1 简单的 insert 方法

下面向 tab_role 表中插入一行数据。SQL 代码如下：

```
<insert id="insertRole" parameterType="Role">
    insert into tab_role(ROLE_NAME, ROLE_DESP)
    values (#{name}, #{description})
</insert>
```

- id 用于标识这条 SQL 语句，它是唯一的。
- parameterType 表示传递的参数类型。
- SQL 中的参数 name、description 要和 Role 实体类中的属性名一致。

## 4.3.2 主键回填

在 4.3.1 节中，我们并没有插入 id 列，因为 tab_role 表设置了主键自动增长。在插入数据成功后，很多时候我们需要获取插入成功后的id，用于其他业务的关联操作。在MyBatis中有两种方法可以解决这个问题，下面分别进行讲解。

### 1. 使用 useGeneratedKeys 回填主键

SQL 代码如下所示：

```
<insert id="insertRole" parameterType="Role" useGeneratedKeys="true" keyProperty="id">
    insert into tab_role(ROLE_NAME, ROLE_DESP)
    values (#{name}, #{description})
</insert>
```

对代码中的新元素解释如下：

- keyProperty：指定 POJO 中哪个属性字段是主键。
- useGeneratedKeys：这个主键是否使用数据库内置生成策略。

插入测试代码，如下所示：

```
@Test
public void insertRole() {
    SqlSession session = null;
    try {
        SqlSessionFactory sqlSessionFactory = SqlSessionFactoryWithXmlUtil.getSqlSessionFactory();
        session = sqlSessionFactory.openSession();
        // 使用 SqlSession 创建 mapper 接口的代理对象
        RoleMapper mapper = session.getMapper(RoleMapper.class);
        Role role = new Role();
        role.setName("项目经理");
        role.setDescription("负责管理项目");
        System.out.println("插入前："+role);
        int i = mapper.insertRole(role);
        session.commit();
        System.out.println("插入后："+role);
    } finally {
        if (session != null) {
            session.close();
        }
    }
}
```

执行完控制台显示部分截图，如图 4-1 所示。

```
插入前：Role(id=null, name=项目经理, description=负责管理项目)
2020-05-27 11:20:36,511 [main] DEBUG [org.apache.ibatis.transaction.jdbc.JdbcTransaction] - Opening JDBC Connection
2020-05-27 11:20:37,751 [main] INFO  [com.alibaba.druid.pool.DruidDataSource] - {dataSource-1} inited
2020-05-27 11:20:37,769 [main] DEBUG [org.apache.ibatis.transaction.jdbc.JdbcTransaction] - Setting autocommit to false on JDBC
 Connection [com.mysql.cj.jdbc.ConnectionImpl@62150f9e]
2020-05-27 11:20:37,812 [main] DEBUG [com.bailiban.mapper.RoleMapper.insertRole] - ==>  Preparing: insert into tab_role(ROLE_NAME,
  ROLE_DESP) values (?, ?)
2020-05-27 11:20:38,164 [main] DEBUG [com.bailiban.mapper.RoleMapper.insertRole] - ==> Parameters: 项目经理(String), 负责管理项目
 (String)
2020-05-27 11:20:38,178 [main] DEBUG [com.bailiban.mapper.RoleMapper.insertRole] - <==    Updates: 1
2020-05-27 11:20:38,256 [main] DEBUG [org.apache.ibatis.transaction.jdbc.JdbcTransaction] - Committing JDBC Connection [com.mysql
 .cj.jdbc.ConnectionImpl@62150f9e]
插入后：Role(id=12, name=项目经理, description=负责管理项目)
2020-05-27 11:20:38,261 [main] DEBUG [org.apache.ibatis.transaction.jdbc.JdbcTransaction] - Resetting autocommit to true on JDBC
 Connection [com.mysql.cj.jdbc.ConnectionImpl@62150f9e]
2020-05-27 11:20:38,262 [main] DEBUG [org.apache.ibatis.transaction.jdbc.JdbcTransaction] - Closing JDBC Connection [com.mysql.cj
 .jdbc.ConnectionImpl@62150f9e]
```

图 4-1

从输出结果可以看出插入前 Role 的 id=null，插入成功后，id=12，这样就实现了主键回填。使用这种方式很简单，但是实际工作中有时候主键的生成规则并不是这么简单，比如表 tab_role 并未设置成主键自增。

我们的要求是：如果 tab_role 表没有数据，就设置主键为 1，否则就取主键最大值加 2，来作为新的主键。这个时候就要用到自定义规则生成主键。

### 2. 使用自定义规则生成主键

SQL 代码如下所示：

```
<insert id="insertRole" parameterType="Role">
    <selectKey keyProperty="id" resultType="int" order="BEFORE">
        select if(max(id) is null,1,max(id)+2) from tab_role
    </selectKey>
    insert into tab_role(id,ROLE_NAME, ROLE_DESP)
    values (#{id},#{name}, #{description})
</insert>
```

使用自定义规则生成主键，也就是使用<selectKey>元素，keyProperty 属性指定 POJO 中哪个属性字段是主键，resultType 指定返回数据类型，order 的可选值为 BEFORE 和 AFTER，分别表示是在插入语句之前执行<selectKey>中的 SQL 语句还是在之后执行。

综上所述，可以发现使用 selectKey 方式非常灵活，生成策略完全可以定制化。例如当使用 MySql 的主键自增方式时，也可以完成 4.3.2 节中的主键回填方式。SQL 代码如下：

```
<insert id="insertRole" parameterType="Role">
    <selectKey keyProperty="id" resultType="int" order="AFTER">
        select LAST_INSERT_ID()
    </selectKey>
    insert into tab_role(ROLE_NAME, ROLE_DESP)
    values (#{name}, #{description})
</insert>
```

上面代码中的 order 值为 AFTER，那是因为在 MySql 数据库中，当前记录的主键值在 insert 语句执行成功后才能获取到。如果是 Oracle 数据库，order 的值要设置为 BEFORE，因为 Oracle 中需要先从序列获取值，然后将值作为主键插入到数据库中。

## 4.4 update 元素和 delete 元素

update 元素和 delete 元素比较简单，所以这里就放一起讲解。它们的元素属性和 insert 基本相同，执行完也会返回一个整数，表示该 SQL 语句影响了数据库记录的行数。下面看一下 update 和 delete 语句的示例：

```
<update id="updateRole" parameterType="role">
    update tab_role
    set ROLE_NAME=#{name},
        ROLE_DESP=#{description}
    where id = #{id}
</update>
<delete id="deleteRole">
    delete
    from tab_role
    where id = #{id}
</delete>
```

上面代码中的元素含义可以参考前面的 select 和 insert 元素。

## 4.5 sql 元素

sql 元素可以用来定义可重用的 SQL 代码片段，以便在其他语句中使用。参数可以静态地(在加载的时候)确定下来，并且可以在不同的 include 元素中定义不同的参数值。代码如下所示：

```
<resultMap id="roleMap" type="Role">
    <id property="id" column="id"/>
    <result property="name" column="ROLE_NAME"/>
    <result property="description" column="ROLE_DESP"/>
</resultMap>
<sql id="roleColumns">ID , ROLE_NAME , ROLE_DESP</sql>
<select id="findRoleById" parameterType="integer" resultMap="roleMap">
    select
    <include refid="roleColumns"/>
```

```
        from tab_role
        where id = #{id}
    </select>
```

上面代码中通过<sql>元素定义了列名，然后查询语句中通过<include>元素的属性 refid 来引入定义的 sql。

sql 元素还支持变量传递，代码如下所示：

```
<resultMap id="roleMap" type="Role">
    <id property="id" column="id"/>
    <result property="name" column="ROLE_NAME"/>
    <result property="description" column="ROLE_DESP"/>
</resultMap>
<sql id="roleColumns">${alias}.ID , ${alias}.ROLE_NAME , ${alias}.ROLE_DESP</sql>
<select id="findRoleById" parameterType="integer" resultMap="roleMap">
    select
    <include refid="roleColumns">
        <property name="alias" value="tr"/>
    </include>
    from tab_role tr
    where id = #{id}
</select>
```

上面代码中，在 include 元素中定义了一个命名为 alias 的变量，其值是 SQL 中表 tab_role 的别名 tr，这样 sql 元素就可以使用这个变量名了。

## 4.6 resultMap 元素

resultMap 元素是 MyBatis 中最重要最强大的元素。它可以把用户从 90% 的 JDBC ResultSets 数据提取代码中解放出来，并在一些情形下允许用户进行一些 JDBC 不支持的操作。其主要作用是建立 SQL 查询结果字段与实体属性的映射关系信息。

### 4.6.1 resultMap 元素的结构

resultMap 元素的子元素，如下面的代码所示：

```
<resultMap id="" type="">
    <constructor>
        <idArg/>
        <arg/>
    </constructor>
    <id/>
```

```
        <result/>
        <association property=""/>
        <collection property=""/>
        <discriminator javaType="">
            <case value=""/>
        </discriminator>
    </resultMap>
```

其中 constructor 元素用于指定构造方法，如果 POJO 中没有无参构造方法，只有有参构造方法，并且查询结果集中的元素又无法和有参构造方法参数对应(顺序不作要求)，那么此时就会报错：POJO 中无构造方法与结果集对应。此时就可以通过 constructor 元素指定构造方法，注意：这里只是指定构造方法，所以前提是 POJO 中有对应的相同的构造方法。这样就可以解决这种场景下的问题。

实际上只要 POJO 中有无参构造方法就不会出现这种情况，因此使用 constructor 元素的时候并不多见，这里大家作为了解即可。

此外还有 association、collection 和 discriminator 这些元素，它们均涉及级联的问题，比较复杂，具体内容将在 4.7 节中详细讲解。

## 4.6.2 使用 map 存储结果集

在前面内容中查询结果都存储在 POJO 中，其实也可以使用 map 存储，代码如下所示：

```
<select id="findUserById" parameterType="integer" resultType="map">
    select id,username, birthday, sex, address
    from tab_user
    where id = #{id}
</select>
```

接口方法代码如下：

```
Map<String,Object> findUserById(@Param("id") Integer id);
```

使用 map 原则上可以匹配所有结果集，但是并不推荐使用这种方式，因为使用 map 意味着可读性下降，更推荐使用 POJO 方式。

## 4.6.3 使用 POJO 存储结果集

使用 POJO 方式存储在 4.2.4 小节中已有演示，对于简单的可以使用自动映射的情况，使用 resultType 属性即可。对于复杂的映射关系，使用 resultMap 来配置。代码如下：

```
<resultMap id="roleMap" type="Role">
    <id property="id" column="id"/>
    <result property="name" column="ROLE_NAME"/>
    <result property="description" column="ROLE_DESP"/>
</resultMap>
```

上面代码中 type 代表需要映射的 POJO，这里使用了别名，也可以使用全限定名，本节就不过多赘述了。

# 4.7 级联

级联是一个数据库实体的概念。MyBatis 的级联操作主要是针对一对一、一对多的情况而设定的。级联是在 resultMap 标签中配置的。在 MyBatis 中还有一种被称为鉴别器的级联，它是一种可以选择具体实现类的级联。

级联不是必须的。级联的好处是获取关联数据十分便捷，但是级联过多会增加系统的复杂度，同时降低系统的性能，此增彼减，所以当级联的层级超过 3 层时，就不要考虑使用级联了，因为这样会造成多个对象的关联，导致系统的耦合、复杂和难以维护。在现实的使用过程中，要根据实际情况判断是否需要使用级联。

## 4.7.1 MyBatis 中的级联

MyBatis 的级联分为 3 种：

- 鉴别器(discriminator)：它是一个根据某些条件决定采用具体实现类的级联，比如根据角色是否可用来决定是否展示权限。
- 一对一(association)：比如公民和身份证，学生和学号都是一种一对一的级联关系。
- 一对多(collection)：比如用户和订单，部门和员工都是一对多的级联关系。

对于多对多级联关系，在 MyBatis 中是没有专门的元素来表示的。这也比较好理解，因为多对多关系，比如用户和角色关系，在实际工作中都是拆分成两个一对多来关联。

为了更好地理解级联，下面新建一个 Maven 项目 mybatis-demo3，其结构和 pom 文件同项目 mybatis-demo2，其中用到的级联数据模型如图 4-2 所示。

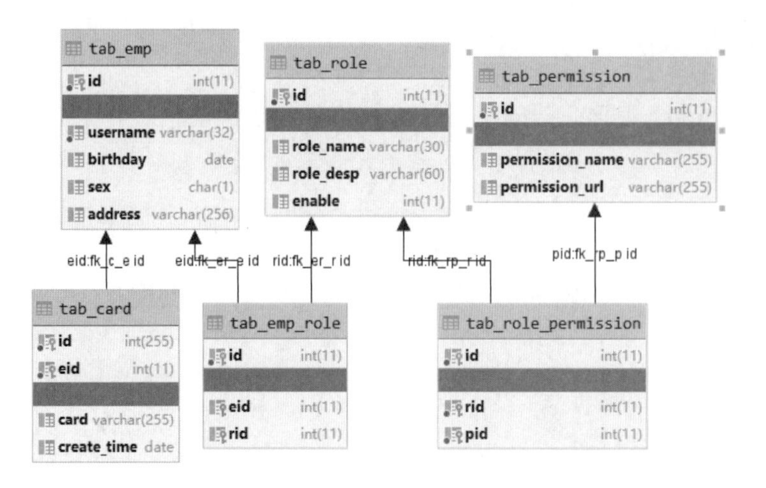


图 4-2

对该用户级联模型解释如下：

- 该模型以员工表为核心。
- 员工和工号信息表是一对一的级联关系。
- 员工和角色是多对多的关系，因此抽离出一个中间表 tab_emp_role，然后转换成两个一对多的级联关系。
- 角色和权限是多对多的关系，同理抽离出一个中间表 tab_role_permission，然后转换成两个一对多的级联关系。

建表的 SQL 语句代码如下所示：

```
/*
Navicat MySQL Data Transfer

Source Server: zhulang
Source Server Version : 50727
Source Host: localhost:3306
Source Database: mybatis-demo3

Target Server Type: MYSQL
Target Server Version : 50727
File Encoding: 65001

Date: 2020-06-04 17:14:54
*/

SET FOREIGN_KEY_CHECKS=0;

-- ----------------------------
-- Table structure for tab_card
```

```
-- ----------------------------
DROP TABLE IF EXISTS `tab_card`;
CREATE TABLE `tab_card` (
  `id` int(255) NOT NULL AUTO_INCREMENT,
  `card` varchar(255) DEFAULT NULL COMMENT '工号',
  `create_time` date DEFAULT NULL COMMENT '创建时间',
  `eid` int(11) NOT NULL,
  PRIMARY KEY (`id`,`eid`),
  UNIQUE KEY `fk_c_e` (`eid`) USING BTREE,
  CONSTRAINT `fk_c_e` FOREIGN KEY (`eid`) REFERENCES `tab_emp` (`id`) ON DELETE
         CASCADE ON UPDATE CASCADE
) ENGINE=InnoDB AUTO_INCREMENT=6 DEFAULT CHARSET=utf8mb4;

-- ----------------------------
-- Records of tab_card
-- ----------------------------
INSERT INTO `tab_card` VALUES ('1', 'a0001', '2020-05-20', '1');
INSERT INTO `tab_card` VALUES ('2', 'a0002', '2020-05-12', '2');
INSERT INTO `tab_card` VALUES ('3', 'a0003', '2020-03-12', '3');
INSERT INTO `tab_card` VALUES ('4', 'a0004', '2020-05-05', '4');
INSERT INTO `tab_card` VALUES ('5', 'a0005', '2020-05-29', '5');

-- ----------------------------
-- Table structure for tab_emp
-- ----------------------------
DROP TABLE IF EXISTS `tab_emp`;
CREATE TABLE `tab_emp` (
  `id` int(11) NOT NULL AUTO_INCREMENT,
  `username` varchar(32) NOT NULL COMMENT '员工名称',
  `birthday` date DEFAULT NULL COMMENT '生日',
  `sex` char(1) DEFAULT NULL COMMENT '性别',
  `address` varchar(256) DEFAULT NULL COMMENT '地址',
  PRIMARY KEY (`id`),
  KEY `id` (`id`)
) ENGINE=InnoDB AUTO_INCREMENT=6 DEFAULT CHARSET=utf8;

-- ----------------------------
-- Records of tab_emp
-- ----------------------------
INSERT INTO `tab_emp` VALUES ('1', '维亚', '1999-10-25', '女', '浙江杭州滨江新园区');
INSERT INTO `tab_emp` VALUES ('2', '阿伦', '1987-10-25', '男', '上海浦东新区');
INSERT INTO `tab_emp` VALUES ('3', '玥玥', '2013-01-15', '女', '广东深圳南山区');
INSERT INTO `tab_emp` VALUES ('4', '张良', '2009-10-25', '男', '湖北武汉东湖高新');
INSERT INTO `tab_emp` VALUES ('5', '宋宝', '2020-05-13', '男', '湖北武汉洪山区');

-- ----------------------------
-- Table structure for tab_emp_role
```

```
-- ----------------------------
DROP TABLE IF EXISTS `tab_emp_role`;
CREATE TABLE `tab_emp_role` (
  `id` int(11) NOT NULL AUTO_INCREMENT,
  `eid` int(11) DEFAULT NULL COMMENT '员工 id',
  `rid` int(11) DEFAULT NULL COMMENT '角色 id',
  PRIMARY KEY (`id`),
  KEY `fk_er_r` (`rid`),
  KEY `fk_er_e` (`eid`),
  CONSTRAINT `fk_er_e` FOREIGN KEY (`eid`) REFERENCES `tab_emp` (`id`) ON DELETE
CASCADE ON UPDATE CASCADE,
  CONSTRAINT `fk_er_r` FOREIGN KEY (`rid`) REFERENCES `tab_role` (`id`) ON DELETE
CASCADE ON UPDATE CASCADE
) ENGINE=InnoDB AUTO_INCREMENT=8 DEFAULT CHARSET=utf8mb4;

-- ----------------------------
-- Records of tab_emp_role
-- ----------------------------
INSERT INTO `tab_emp_role` VALUES ('1', '1', '1');
INSERT INTO `tab_emp_role` VALUES ('2', '2', '2');
INSERT INTO `tab_emp_role` VALUES ('3', '3', '3');
INSERT INTO `tab_emp_role` VALUES ('4', '4', '4');
INSERT INTO `tab_emp_role` VALUES ('5', '4', '11');
INSERT INTO `tab_emp_role` VALUES ('6', '5', '3');
INSERT INTO `tab_emp_role` VALUES ('7', '5', '8');

-- ----------------------------
-- Table structure for tab_permission
-- ----------------------------
DROP TABLE IF EXISTS `tab_permission`;
CREATE TABLE `tab_permission` (
  `id` int(11) NOT NULL AUTO_INCREMENT,
  `permission_name` varchar(255) DEFAULT NULL COMMENT '权限名称',
  `permission_url` varchar(255) DEFAULT NULL COMMENT '可访问 url',
  PRIMARY KEY (`id`)
) ENGINE=InnoDB AUTO_INCREMENT=7 DEFAULT CHARSET=utf8mb4;

-- ----------------------------
-- Records of tab_permission
-- ----------------------------
INSERT INTO `tab_permission` VALUES ('1', '查询所有员工', '/emp/findAll');
INSERT INTO `tab_permission` VALUES ('2', '新建员工', '/emp/save');
INSERT INTO `tab_permission` VALUES ('3', '查询员工详情', '/emp/findById');
INSERT INTO `tab_permission` VALUES ('4', '任务查询', '/task/findAll');
INSERT INTO `tab_permission` VALUES ('5', '新建任务', '/task/save');
INSERT INTO `tab_permission` VALUES ('6', '查询进度', '/task/progress');
```

```
-- ----------------------------
-- Table structure for tab_role
-- ----------------------------
DROP TABLE IF EXISTS `tab_role`;
CREATE TABLE `tab_role` (
  `id` int(11) NOT NULL AUTO_INCREMENT COMMENT '编号',
  `role_name` varchar(30) DEFAULT NULL COMMENT '角色名称',
  `role_desp` varchar(60) DEFAULT NULL COMMENT '角色描述',
  `enable` int(11) DEFAULT '1' COMMENT '0：不可用，1：可用',
  PRIMARY KEY (`id`)
) ENGINE=InnoDB AUTO_INCREMENT=13 DEFAULT CHARSET=utf8;

-- ----------------------------
-- Records of tab_role
-- ----------------------------
INSERT INTO `tab_role` VALUES ('1', '董事长', '企业老大', '1');
INSERT INTO `tab_role` VALUES ('2', '总经理', '管理整个公司', '1');
INSERT INTO `tab_role` VALUES ('3', '副总经理', '分管其中一个片区', '1');
INSERT INTO `tab_role` VALUES ('4', '人力 hr', '负责招聘员工组织管理', '0');
INSERT INTO `tab_role` VALUES ('5', '会计', '负责财务', '1');
INSERT INTO `tab_role` VALUES ('6', '研发经理', '负责产品研发', '1');
INSERT INTO `tab_role` VALUES ('7', '市场总监', '负责市场', '1');
INSERT INTO `tab_role` VALUES ('8', '产品经理', '产品的设计', '1');
INSERT INTO `tab_role` VALUES ('9', '运营总监', '产品的运营维护', '1');
INSERT INTO `tab_role` VALUES ('10', '设计总监', '平面设计', '1');
INSERT INTO `tab_role` VALUES ('11', '职业规划师', '负责职业规划', '1');
INSERT INTO `tab_role` VALUES ('12', '项目经理', '负责管理项目', '1');

-- ----------------------------
-- Table structure for tab_role_permission
-- ----------------------------
DROP TABLE IF EXISTS `tab_role_permission`;
CREATE TABLE `tab_role_permission` (
  `id` int(11) NOT NULL AUTO_INCREMENT,
  `rid` int(11) NOT NULL COMMENT '角色 id',
  `pid` int(11) NOT NULL COMMENT '权限 id',
  PRIMARY KEY (`id`),
  KEY `fk_rp_r` (`rid`),
  KEY `fk_rp_p` (`pid`),
  CONSTRAINT `fk_rp_p` FOREIGN KEY (`pid`) REFERENCES `tab_permission` (`id`) ON DELETE
CASCADE ON UPDATE CASCADE,
  CONSTRAINT `fk_rp_r` FOREIGN KEY (`rid`) REFERENCES `tab_role` (`id`) ON DELETE
CASCADE ON UPDATE CASCADE
) ENGINE=InnoDB AUTO_INCREMENT=15 DEFAULT CHARSET=utf8mb4;

-- ----------------------------
-- Records of tab_role_permission
```

```
-- ----------------------------
INSERT INTO `tab_role_permission` VALUES ('1', '1', '1');
INSERT INTO `tab_role_permission` VALUES ('2', '1', '3');
INSERT INTO `tab_role_permission` VALUES ('3', '1', '4');
INSERT INTO `tab_role_permission` VALUES ('4', '1', '6');
INSERT INTO `tab_role_permission` VALUES ('5', '4', '2');
INSERT INTO `tab_role_permission` VALUES ('6', '2', '1');
INSERT INTO `tab_role_permission` VALUES ('7', '3', '1');
INSERT INTO `tab_role_permission` VALUES ('8', '5', '1');
INSERT INTO `tab_role_permission` VALUES ('9', '6', '1');
INSERT INTO `tab_role_permission` VALUES ('10', '7', '1');
INSERT INTO `tab_role_permission` VALUES ('11', '8', '1');
INSERT INTO `tab_role_permission` VALUES ('12', '9', '1');
INSERT INTO `tab_role_permission` VALUES ('13', '10', '1');
INSERT INTO `tab_role_permission` VALUES ('14', '11', '6');
INSERT INTO `tab_role_permission` VALUES ('15', '12', '1');
```

## 4.7.2 建立 POJO

根据数据模型，建立对应的 POJO，这里需要创建 4 个实体类，分别为 Card、Emp、Role 和 Permission，放到 com.bailiban.entity 包下。代码如下：

### 1.Card(工号类)

```
@Data
public class Card implements Serializable {
    private static final long serialVersionUID = -6434310878590492 3L;

    private Integer id;
    /**
     * 工号
     */
    private String card;
    /**
     * 创建时间
     */
    private Date createTime;
    /**
     * 一对一：员工
     */
    private Emp emp;
}
```

Card 类中，有一个成员变量是 Emp 员工类，建立起员工工号和员工之间一对一的级联关系。

注意：在表 tab_card 中有一个外键 eid 关联表 tab_emp 中的主键 id，但是在 Card 实体类中，我们并没有加入对应的 eid 属性，因为外键对实体类而言并没有特殊的含义。

### 2. Emp(员工类)

```
@Data
public class Emp implements Serializable {
    private static final long serialVersionUID = 826935895632451598L;

    private Integer id;
    /**
     * 员工名称
     */
    private String username;
    /**
     * 生日
     */
    private Date birthday;
    /**
     * 性别
     */
    private String sex;
    /**
     * 地址
     */
    private String address;
    /**
     * 一对多：角色
     */
    private List<Role> roles;
}
```

从员工来讲，Emp 和 Role 是一对多的级联关系，所以在 Emp 类中加入集合 roles 属性来建立关系。

Emp 和 Card 是一对一的级联关系，可以在 Emp 类中加入 Card 属性。因为在 Card 类中已经添加了一对一的级联关系，所以在 Emp 表中就没有再添加。

### 3. Role(角色类)

```
@Data
public class Role implements Serializable {
    private static final long serialVersionUID = 705499652105635382L;
    /**
     * 编号
     */
    private Integer id;
    /**
     * 角色名称
```

```
     */
    private String roleName;
    /**
     * 角色描述
     */
    private String roleDesp;
    /**
     * 角色是否可用，1：可用，0：不可用
     */
    private int enable;
    /**
     * 一对多：员工
     */
    private List<Emp> emps;
    /**
     * 一对多：权限
     */
    private List<Permission> permissions;
}
```

同上，对于角色而言，Role 和 Emp 是一对多的级联关系，Role 和 Permission 也是一对多的级联关系，在 Role 类中加入集合 emps 属性和集合 permissions 属性来建立关系。

### 4. Permission(权限类)

```
@Data
public class Permission implements Serializable {
    private static final long serialVersionUID = 5228717250349307491L;

    private Integer id;
    /**
     * 权限名称
     */
    private String permissionName;
    /**
     * 可访问 url
     */
    private String permissionUrl;
    /**
     * 一对多：角色
     */
    private List<Role> roles;
}
```

对于权限而言，Permission 和 Role 是一对多的级联关系，在 Permission 类中加入集合 roles 属性来建立关系。

## 4.7.3 配置映射文件

创建四个映射文件，分别为 CardMapper.xml、EmpMapper.xml、RoleMapper.xml 和 PermissionMapper.xml，进行基础的配置，写有一个根据 id 查询的 SQL，代码如下所示：

### 1. CardMapper.xml

```
<?xml version="1.0" encoding="UTF-8"?>
<!DOCTYPE mapper PUBLIC "-//mybatis.org//DTD Mapper 3.0//EN"
            "http://mybatis.org/dtd/mybatis-3-mapper.dtd">
<mapper namespace="com.bailiban.mapper.CardMapper">

    <resultMap type="com.bailiban.entity.Card" id="CardMap">
        <id property="id" column="cid"/>
        <result property="card" column="card"/>
        <result property="createTime" column="create_time"/>
    </resultMap>

    <!--查询单个-->
    <select id="queryById" resultMap="CardMap">
        select id cid,
                card,
                create_time
        from tab_card
        where id = #{id}
    </select>
</mapper>
```

### 2. EmpMapper.xml

```
<?xml version="1.0" encoding="UTF-8"?>
<!DOCTYPE mapper PUBLIC "-//mybatis.org//DTD Mapper 3.0//EN"
            "http://mybatis.org/dtd/mybatis-3-mapper.dtd">
<mapper namespace="com.bailiban.mapper.EmpMapper">

    <resultMap type="com.bailiban.entity.Emp" id="EmpMap">
        <id property="id" column="eid"/>
        <result property="username" column="username"/>
        <result property="birthday" column="birthday"/>
        <result property="sex" column="sex"/>
        <result property="address" column="address"/>
    </resultMap>
    <!--查询单个-->
    <select id="queryById" resultMap="EmpMap">
        select id eid,
                username,
                birthday,
```

```
                    sex,
                    address
            from tab_emp
            where id = #{id}
    </select>
</mapper>
```

### 3. RoleMapper.xml

```
<?xml version="1.0" encoding="UTF-8"?>
<!DOCTYPE mapper PUBLIC "-//mybatis.org//DTD Mapper 3.0//EN"
        "http://mybatis.org/dtd/mybatis-3-mapper.dtd">
<mapper namespace="com.bailiban.mapper.RoleMapper">

    <resultMap type="com.bailiban.entity.Role" id="RoleMap">
        <id property="id" column="rid"/>
        <result property="roleName" column="role_name"/>
        <result property="roleDesp" column="role_desp"/>
        <result property="enable" column="enable"/>
    </resultMap>
    <!--查询单个-->
    <select id="queryById" resultMap="RoleMap">
        select id rid,
                role_name,
                role_desp,
                enable
        from tab_role
        where id = #{id}
    </select>
</mapper>
```

### 4. PermissionMapper.xml

```
<?xml version="1.0" encoding="UTF-8"?>
<!DOCTYPE mapper PUBLIC "-//mybatis.org//DTD Mapper 3.0//EN"
        "http://mybatis.org/dtd/mybatis-3-mapper.dtd">
<mapper namespace="com.bailiban.mapper.PermissionMapper">

    <resultMap type="com.bailiban.entity.Permission" id="PermissionMap">
        <id property="id" column="pid"/>
        <result property="permissionName" column="permission_name"/>
        <result property="permissionUrl" column="permission_url"/>
    </resultMap>

    <!--查询单个-->
    <select id="queryById" resultMap="PermissionMap">
        select id pid,
                permission_name,
                permission_url
```

```
            from tab_permission
            where id = #{id}
        </select>
    </mapper>
```

完成以上映射文件的配置，还要编写对应的接口方法，由于其十分简单，这里就不进行展示。

### 4.7.4 association 标签配置一对一级联

员工工号和员工是一对一的级联关系，我们在查询员工工号信息的时候，同时级联查询出对应的员工信息，在 CardMapper.xml 中加入如下代码：

```
<resultMap id="Card_EmpMap" type="Card">
    <id property="id" column="cid"/>
    <result property="card" column="card"/>
    <result property="createTime" column="create_time"/>
    <association property="emp" javaType="Emp">
        <id property="id" column="eid"/>
        <result property="username" column="username"/>
        <result property="birthday" column="birthday"/>
        <result property="sex" column="sex"/>
        <result property="address" column="address"/>
    </association>
</resultMap>
<select id="queryCardEmpById" resultMap="Card_EmpMap">
    select tc.id cid,
            tc.card,
            tc.create_time,
            te.id eid,
            te.username,
            te.birthday,
            te.sex,
            te.address
    from tab_card tc
            join tab_emp te on tc.eid = te.id
    where tc.id = #{id}
</select>
```

上面代码中，SQL 部分使用了内连接关联查询出工号和对应员工的信息，查询的结果集通过 resultMap 进行映射，<resultMap>元素内<association>元素外的部分通过<id>和<result>元素配置了 Card 类的属性和 SQL 列的自动映射关系，同样的在<association>元素内也通过<id>和<result>元素配置了 Emp 类的属性和 SQL 列的映射关系。

关于 association 标签包含的属性解释如下：

- property：对应实体类中的属性名，必填项。
- javaType：属性对应的 Java 类型。
- resultMap：可以直接使用现有的 resultMap，而不需要在这里配置。

除了这些属性外，还有其他属性，此处不做介绍。

其实上面代码中的 resultMap 部分可以进一步简化，因为里面配置的关于 Card 和 Emp 的基本映射都已经配置好了，无须重复配置。所以简化后的代码如下所示：

```
<!-- 方式二：一对一级联-->
<resultMap id="Card_EmpMap" type="Card" extends="CardMap">
    <association property="emp" javaType="Emp"
    resultMap="com.bailiban.mapper.EmpMapper.EmpMap"/>
</resultMap>
```

上面代码中，使用 extends 属性来继承定义好的 id 为 CardMap 的 resultMap。association 标签中直接使用 resultMap 属性引用 EmpMapper.xml 中配置好的 id 为 EmpMap 的 resultMap。这样我们的配置就变得简单了许多。

association 标签中除了可以使用 resultMap 属性来引用映射结果集，还可以使用 select 属性来实现。我们将会在 4.7.5 节进行使用和讲解。

下面在测试类中进行测试，测试代码如下：

```
@Test
public void queryCardEmpById() {
    SqlSession sqlSession = null;
    try {
        SqlSessionFactory sqlSessionFactory = SqlSessionFactoryWithXmlUtil.getSqlSessionFactory();
        // 生成 sqlSession
        sqlSession = sqlSessionFactory.openSession();
        // 使用 SqlSession 创建 mapper 接口的代理对象
        CardMapper mapper = sqlSession.getMapper(CardMapper.class);
        Card card = mapper.queryCardEmpById(1);
        System.out.println(card);
    } finally {
        if (sqlSession != null) {
            sqlSession.close();
        }
    }
}
```

控制台输出的结果如图 4-3 所示。

```
2020-06-04 17:26:07,222 [main] DEBUG [com.bailiban.mapper.CardMapper.queryCardEmpById] - ==>  Preparing: select tc.id cid,
 tc.card, tc.create_time, te.id eid, te.username, te.birthday, te.sex, te.address from tab_card tc join tab_emp te on tc.eid =
 te.id where tc.id = ?
2020-06-04 17:26:07,422 [main] DEBUG [com.bailiban.mapper.CardMapper.queryCardEmpById] - ==> Parameters: 1(Integer)
2020-06-04 17:26:07,504 [main] DEBUG [com.bailiban.mapper.CardMapper.queryCardEmpById] - <==      Total: 1
Card(id=1, card=a0001, createTime=Wed May 20 00:00:00 CST 2020, emp=Emp(id=1, username=维亚, birthday=Mon Oct 25 00:00:00 CST
 1999, sex=女, address=浙江杭州滨江新园区, roles=null))
2020-06-04 17:26:07,506 [main] DEBUG [org.apache.ibatis.transaction.jdbc.JdbcTransaction] - Resetting autocommit to true on JDBC
 Connection [com.mysql.cj.jdbc.ConnectionImpl@5a63f509]
```

图 4-3

## 4.7.5 collection 标签配置一对多级联

对于员工而言，员工和角色是一对多级联关系，在查询员工信息时，级联查询出其对应的所有角色信息。在 RoleMapper.xml 中加入如下代码：

```
<select id="findRolesByEid" resultMap="RoleMap">
    select tr.id rid,
           tr.role_name,
           tr.role_desp,
           tr.enable
    from tab_role tr,
         tab_emp_role ter
    where tr.id = ter.rid
      and ter.eid = #{eid}
</select>
```

上面代码中 SQL 部分使用了隐式内连接关联查询出所有的角色信息，查询的结果集通过 id 为 RoleMap 的 resultMap 进行映射。

再在 EmpMapper.xml 中加入如下代码：

```
<resultMap id="Emp_RoleMap" type="Emp" extends="EmpMap">
    <collection property="roles" column="eid"
      select="com.bailiban.mapper.RoleMapper.findRolesByEid" ofType="Role"/>
</resultMap>
<select id="findEmpRoleById" resultMap="Emp_RoleMap">
    select e.id eid,
           e.username,
           e.birthday,
           e.sex,
           e.address
    from tab_emp e
    where id = #{id}
</select>
```

上面代码中在 collection 标签元素中直接使用 select 属性引用 RoleMapper.xml 中配置好的 id 为 findRolesByEid 的 select 查询语句。collection 标签中的 column 属性表示传递参数的列，ofType 属性指定映射到集合中的 POJO 对象类型，可以省略。

当然 collection 标签自身也含有 resultMap 属性，类似于 association 标签，也可以使用

resultMap 完成一对多结果集的映射，这里就不再进行演示。

在测试类中进行测试，测试代码如下：

```
@Test
public void findEmpRoleById() {
    SqlSession sqlSession = null;
    try {
        SqlSessionFactory sqlSessionFactory = SqlSessionFactoryWithXmlUtil.getSqlSessionFactory();
        // 生成 sqlSession
        sqlSession = sqlSessionFactory.openSession();
        // 使用 SqlSession 创建 mapper 接口的代理对象
        EmpMapper mapper = sqlSession.getMapper(EmpMapper.class);
        Emp emp = mapper.findEmpRoleById(4);
        System.out.println(emp);
    } finally {
        if (sqlSession != null) {
            sqlSession.close();
        }
    }
}
```

控制台输出结果，如图 4-4 所示。

```
2020-06-04 18:05:01,854 [main] DEBUG [com.bailiban.mapper.EmpMapper.findEmpRoleById] - ==>  Preparing: select e.id eid, e
 .username, e.birthday, e.sex, e.address from tab_emp e where id = ?
2020-06-04 18:05:02,029 [main] DEBUG [com.bailiban.mapper.EmpMapper.findEmpRoleById] - ==> Parameters: 4(Integer)
2020-06-04 18:05:02,184 [main] DEBUG [com.bailiban.mapper.RoleMapper.findRolesByEid] - ====>  Preparing: select tr.id rid,
 tr.role_name, tr.role_desp, tr.enable from tab_role tr, tab_emp_role ter where tr.id = ter.rid and ter.eid = ?
2020-06-04 18:05:02,184 [main] DEBUG [com.bailiban.mapper.RoleMapper.findRolesByEid] - ====> Parameters: 4(Integer)
2020-06-04 18:05:02,187 [main] DEBUG [com.bailiban.mapper.RoleMapper.findRolesByEid] - <====      Total: 2
2020-06-04 18:05:02,190 [main] DEBUG [com.bailiban.mapper.EmpMapper.findEmpRoleById] - <==      Total: 1
Emp(id=4, username=张良, birthday=Sun Oct 25 00:00:00 CST 2009, sex=男, address=湖北武汉东湖高新, roles=[Role(id=4, roleName=人力hr,
  roleDesp=负责招聘员工组织管理, enable=0, emps=null, permissions=null), Role(id=11, roleName=职业规划师, roleDesp=负责职业规划,
 enable=1, emps=null, permissions=null)])
2020-06-04 18:05:02,191 [main] DEBUG [org.apache.ibatis.transaction.jdbc.JdbcTransaction] - Resetting autocommit to true on JDBC
 Connection [com.mysql.cj.jdbc.ConnectionImpl@5a63f509]
```

图 4-4

## 4.7.6 discriminator 标签配置级联

鉴别器(discriminator)标签是一个根据某些条件决定采用具体实现类级联的方案。理解起来非常容易，因为它很像 Java 语言中的 switch case 语句。

对 discriminator 标签包含的常用属性解释如下：

- column：用于设置要进行鉴别比较值的列。
- javaType：属性对应的 Java 类型。

discriminator 标签下面可以有一个或多个 case 标签，case 标签包含以下三个属性。

- value：该值为 discriminator 指定 column 用来匹配的值。

- resultMap：当 column 的值和 value 的值匹配时，可以配置使用 resultMap 指定的映射，resultMap 优先级高于 resultType。
- resultType：当 column 的值和 value 的值匹配时，可以配置使用 resultType 指定的映射。

在 RoleMapper.xml 中加入如下代码：

```
<resultMap id="Role_PermissionMap" type="com.bailiban.entity.Role" extends="RoleMap">
    <collection property="permissions" ofType="Permission"
            resultMap="com.bailiban.mapper.PermissionMapper.PermissionMap"/>
</resultMap>
<resultMap id="baseMap" type="Role">
    <discriminator javaType="int" column="enable">
        <case value="0" resultMap="RoleMap"/>
        <case value="1" resultMap="Role_PermissionMap"/>
    </discriminator>
</resultMap>
<select id="findRolesPermissionByEid" resultMap="baseMap">
    select tr.id rid,
           tr.role_name,
           tr.role_desp,
           tr.enable,
           tp.id pid,
           tp.permission_name,
           tp.permission_url
    from tab_role tr
            left join tab_role_permission trp on tr.id = trp.rid
            left join tab_permission tp on tp.id = trp.pid
            left join tab_emp_role ter on ter.rid = tr.id
    where ter.eid = #{eid}
</select>
```

表 tab_role 中有一个字段 enable，当其值为 0 时表示不可用，此时就不级联查询出权限信息；当其值为 1 时表示可用，此时就级联查询出权限信息。

再在 EmpMapper.xml 中加入如下代码：

```
<resultMap id="Emp_Role_PermissionMap" type="Emp" extends="EmpMap">
    <collection property="roles" select="com.bailiban.mapper.RoleMapper.findRolesPermissionByEid"
            column="eid"/>
</resultMap>
<select id="findEmpRolePermissionById" resultMap="Emp_Role_PermissionMap">
    select e.id eid,
           e.username,
           e.birthday,
           e.sex,
           e.address
```

```
        from tab_emp e
        where e.id = #{id}
</select>
```

这样就可以查询员工信息同时级联查询出其所有角色信息，并且根据角色是否可用来选择对应的 resultMap 进行权限信息的映射。

在测试类中进行测试，测试代码如下：

```
@Test
public void findEmpRolePermissionById() {
    SqlSession sqlSession = null;
    try {
        SqlSessionFactory sqlSessionFactory = SqlSessionFactoryWithXmlUtil.getSqlSessionFactory();
        //  生成 sqlSession
        sqlSession = sqlSessionFactory.openSession();
        //   使用 SqlSession 创建 mapper 接口的代理对象
        EmpMapper mapper = sqlSession.getMapper(EmpMapper.class);
        Emp emp = mapper.findEmpRolePermissionById(4);
        System.out.println(emp);
    } finally {
        if (sqlSession != null) {
            sqlSession.close();
        }
    }
}
```

控制台输出结果，如图 4-5 所示。

```
2020-06-05 11:36:20,872 [main] DEBUG [com.bailiban.mapper.EmpMapper.findEmpRolePermissionById] - ==>  Preparing: select e.id eid,
 e.username, e.birthday, e.sex, e.address from tab_emp e where e.id = ?
2020-06-05 11:36:21,010 [main] DEBUG [com.bailiban.mapper.EmpMapper.findEmpRolePermissionById] - ==> Parameters: 4(Integer)
2020-06-05 11:36:21,142 [main] DEBUG [com.bailiban.mapper.RoleMapper.findRolesPermissionByEid] - ====>  Preparing: select tr.id
 rid, tr.role_name, tr.role_desp, tr.enable, tp.id pid, tp.permission_name, tp.permission_url from tab_role tr left join
 tab_role_permission trp on tr.id = trp.rid left join tab_permission tp on tp.id = trp.pid left join tab_emp_role ter on ter.rid =
  tr.id where ter.eid = ?
2020-06-05 11:36:21,143 [main] DEBUG [com.bailiban.mapper.RoleMapper.findRolesPermissionByEid] - ====> Parameters: 4(Integer)
2020-06-05 11:36:21,161 [main] DEBUG [com.bailiban.mapper.RoleMapper.findRolesPermissionByEid] - <====      Total: 2
2020-06-05 11:36:21,164 [main] DEBUG [com.bailiban.mapper.EmpMapper.findEmpRolePermissionById] - <==      Total: 1
Emp(id=4, username=张良, birthday=Sun Oct 25 00:00:00 CST 2009, sex=男, address=湖北武汉东湖高新, roles=[Role(id=4, roleName=人力hr,
  roleDesp=负责招聘员工组织管理, enable=0, emps=null, permissions=null), Role(id=11, roleName=职业规划师, roleDesp=负责职业规划,
 enable=1, emps=null, permissions=[Permission(id=6, permissionName=查询进度, permissionUrl=/task/progress, roles=null)])])
2020-06-05 11:36:21,166 [main] DEBUG [org.apache.ibatis.transaction.jdbc.JdbcTransaction] - Resetting autocommit to true on JDBC
 Connection [com.mysql.cj.jdbc.ConnectionImpl@5a63f509]
```

图 4-5

从控制台输出的结果可以看出 id 为 4 的员工有两个角色，其中 id 为 4 的角色由于其 enable=0，所以其权限信息就没有进行映射，而 id 为 11 的角色可用，因此其权限信息自动映射到对应的 POJO 类上。

### 4.7.7 延迟加载

前面使用级联以方便数据的一次性获取，但是有时候又希望级联的数据不要直接查询到，而是在需要用到的时候才取出。例如，在查询员工信息的时候，并不一定需要及时返回其角色信息，在查询角色信息的时候也不一定需要及时返回其权限信息。这个时候就可以使用延迟加载功能。

MyBatis 支持延迟加载，延迟加载其实就是将数据加载时机推迟，在 MyBatis 的 settings 配置中存在两个元素可以配置延迟加载，如表 4-4 所示。

表 4-4

| 配置项 | 描述 | 有效值 | 默认值 |
| --- | --- | --- | --- |
| lazyLoadingEnabled | 延迟加载的全局开关。当开启时，所有关联对象都会延迟加载。在特定关联关系中，可通过设置 fetchType 属性来覆盖该项的开关状态 | true\|false | false |
| aggressiveLazyLoading | 当启用时，对任意延迟属性的调用会使带有延迟加载属性的对象完整加载；反之，则每种属性按需加载 | true\|false | false( 在 3.4.1 及之前的版本中默认为 true) |

在 mybatis-config.xml 中加入如下配置代码：

```
<settings>
    <!--指定 MyBatis 使用 log4j -->
    <setting name="logImpl" value="LOG4J"/>
    <!--开启延迟加载功能-->
    <setting name="lazyLoadingEnabled" value="true"/>
    <!--不进行完整对象的加载，属性按需加载-->
    <setting name="aggressiveLazyLoading" value="false"/>
</settings>
```

在测试类中进行测试，测试代码如下：

```
@Test
public void findEmpRoleById() {
    SqlSession sqlSession = null;
    try {
        SqlSessionFactory sqlSessionFactory = SqlSessionFactoryWithXmlUtil.getSqlSessionFactory();
        // 生成 sqlSession
        sqlSession = sqlSessionFactory.openSession();
        // 使用 SqlSession 创建 mapper 接口的代理对象
        EmpMapper mapper = sqlSession.getMapper(EmpMapper.class);
```

```
            Emp emp = mapper.findEmpRoleById(4);
            System.out.println("-------------打印 emp 对象 username 信息-----------------");
            System.out.println(emp.getUsername());
        } finally {
            if (sqlSession != null) {
                sqlSession.close();
            }
        }
    }
```

控制台输出结果如图 4-6 所示。

```
2020-06-09 10:39:41,788 [main] DEBUG [org.apache.ibatis.transaction.jdbc.JdbcTransaction] - Setting autocommit to false on JDBC
 Connection [com.mysql.cj.jdbc.ConnectionImpl@635eaaf1]
2020-06-09 10:39:41,798 [main] DEBUG [com.bailiban.mapper.EmpMapper.findEmpRoleById] - ==>  Preparing: select e.id eid, e
 .username, e.birthday, e.sex, e.address from tab_emp e where id = ?
2020-06-09 10:39:41,906 [main] DEBUG [com.bailiban.mapper.EmpMapper.findEmpRoleById] - ==> Parameters: 4(Integer)
2020-06-09 10:39:42,150 [main] DEBUG [com.bailiban.mapper.EmpMapper.findEmpRoleById] - <==      Total: 1
-------------打印emp对象username信息-----------------
张良
2020-06-09 10:39:42,182 [main] DEBUG [org.apache.ibatis.transaction.jdbc.JdbcTransaction] - Resetting autocommit to true on JDBC
 Connection [com.mysql.cj.jdbc.ConnectionImpl@635eaaf1]
```

图 4-6

从控制台输出结果可以看出，当只输出 Emp 对象自身信息时，仅发送了一条 SQL 语句，而其关联的 Role 对象信息并未立即进行查询，采取了延迟加载策略。

如果修改 aggressiveLazyLoading 的值为 true 时，再次运行以上测试代码，查看控制台输出结果，如图 4-7 所示。

```
2020-06-09 10:48:31,960 [main] DEBUG [org.apache.ibatis.transaction.jdbc.JdbcTransaction] - Setting autocommit to false on JDBC
 Connection [com.mysql.cj.jdbc.ConnectionImpl@635eaaf1]
2020-06-09 10:48:31,977 [main] DEBUG [com.bailiban.mapper.EmpMapper.findEmpRoleById] - ==>  Preparing: select e.id eid, e
 .username, e.birthday, e.sex, e.address from tab_emp e where id = ?
2020-06-09 10:48:32,193 [main] DEBUG [com.bailiban.mapper.EmpMapper.findEmpRoleById] - ==> Parameters: 4(Integer)
2020-06-09 10:48:32,539 [main] DEBUG [com.bailiban.mapper.RoleMapper.findRolesByEid] - ====>  Preparing: select tr.id rid,
 tr.role_name, tr.role_desp, tr.enable from tab_role tr, tab_emp_role ter where tr.id = ter.rid and ter.eid = ?
2020-06-09 10:48:32,539 [main] DEBUG [com.bailiban.mapper.RoleMapper.findRolesByEid] - ====> Parameters: 4(Integer)
2020-06-09 10:48:32,546 [main] DEBUG [com.bailiban.mapper.RoleMapper.findRolesByEid] - <====      Total: 2
2020-06-09 10:48:32,548 [main] DEBUG [com.bailiban.mapper.EmpMapper.findEmpRoleById] - <==      Total: 1
-------------打印emp对象username信息-----------------
张良
2020-06-09 10:48:32,573 [main] DEBUG [org.apache.ibatis.transaction.jdbc.JdbcTransaction] - Resetting autocommit to true on JDBC
 Connection [com.mysql.cj.jdbc.ConnectionImpl@635eaaf1]
```

图 4-7

从输出的日志图中可以看出此时发送了两条 SQL 语句，一条是查询 Emp 信息，一条是查询关联的 Role 信息。因为 Emp 是带有懒加载属性的对象，因此会完整加载 Emp 对象，即会触发级联查询效果。

有时我们想采用懒加载，但是对于某些数据又希望能级联查询出，此时可以使用 fetchType 属性，fetchType 出现在级联元素(association、collection，注意，discriminator 没有这个属性可配置)中，它存在两个值：

- eager：获得当前 POJO 后立即加载对应的数据。

- lazy：获得当前 POJO 后延迟加载对应的数据。

下面将 lazyLoadingEnabled 设置为 true，aggressiveLazyLoading 设置为 false，对 RoleMapper.xml 中的 id="Emp_RoleMap"的 resultMap 进行修改，代码如下所示：

```
<resultMap id="Emp_RoleMap" type="Emp" extends="EmpMap">
    <collection property="roles" fetchType="eager" column="eid"
                select="com.bailiban.mapper.RoleMapper.findRolesByEid"/>
</resultMap>
```

再次运行方法 findEmpRoleById()测试代码，查看控制台显示的结果如图 4-8 所示。

```
2020-06-09 12:33:33,628 [main] DEBUG [org.apache.ibatis.transaction.jdbc.JdbcTransaction] - Setting autocommit to false on JDBC
 Connection [com.mysql.cj.jdbc.ConnectionImpl@635eaaf1]
2020-06-09 12:33:33,637 [main] DEBUG [com.bailiban.mapper.EmpMapper.findEmpRoleById] - ==>  Preparing: select e.id eid, e
 .username, e.birthday, e.sex, e.address from tab_emp e where id = ?
2020-06-09 12:33:33,725 [main] DEBUG [com.bailiban.mapper.EmpMapper.findEmpRoleById] - ==> Parameters: 4(Integer)
2020-06-09 12:33:33,817 [main] DEBUG [com.bailiban.mapper.RoleMapper.findRolesByEid] - ====>  Preparing: select tr.id rid,
 tr.role_name, tr.role_desp, tr.enable from tab_role tr, tab_emp_role ter where tr.id = ter.rid and ter.eid = ?
2020-06-09 12:33:33,818 [main] DEBUG [com.bailiban.mapper.RoleMapper.findRolesByEid] - ====> Parameters: 4(Integer)
2020-06-09 12:33:33,821 [main] DEBUG [com.bailiban.mapper.RoleMapper.findRolesByEid] - <====      Total: 2
2020-06-09 12:33:33,823 [main] DEBUG [com.bailiban.mapper.EmpMapper.findEmpRoleById] - <==      Total: 1
-------------打印emp对象username信息----------------
张良
2020-06-09 12:33:33,844 [main] DEBUG [org.apache.ibatis.transaction.jdbc.JdbcTransaction] - Resetting autocommit to true on JDBC
 Connection [com.mysql.cj.jdbc.ConnectionImpl@635eaaf1]
```

图 4-8

这时 Emp 对象相关联的 Role 信息也都执行了关联查询，fetchType 属性会忽略全局配置项 lazyLoadingEnabled 和 aggressiveLazyLoading。

## 4.8 缓存配置

使用缓存在一定程度上可以大大提高数据的读取性能，尤其是对于查询量大和使用频繁的数据，其缓存命中率越高，越能体现出使用缓存的作用。MyBatis 作为持久层框架，也提供了使用缓存的功能。

MyBatis 中缓存分为一级缓存和二级缓存，默认情况下一级缓存是开启的，而且是不能关闭的。因此我们常说的 MyBatis 缓存，都是指二级缓存。下面分别进行讲解。

### 4.8.1 一级缓存

一级缓存是在 SqlSession 上的缓存，下面通过一个简单的示例来看看 MyBatis 一级缓存如何起作用。

```
public class CacheTest {
    @Test
```

```
        public void findEmpById() {
            SqlSession sqlSession = null;
            SqlSession sqlSession1 = null;
            try {
                SqlSessionFactory sqlSessionFactory =
                        SqlSessionFactoryWithXmlUtil.getSqlSessionFactory();
                // 生成 sqlSession
                sqlSession = sqlSessionFactory.openSession();
                // 使用 SqlSession 创建 mapper 接口的代理对象
                EmpMapper mapper = sqlSession.getMapper(EmpMapper.class);
                Emp emp1 = mapper.queryById(4);
                System.out.println("第一次查询结果：" + emp1);
                emp1.setUsername("测试员工名");
                // 修改操作并未提交，未修改数据库
                System.out.println("--------------分割线----------------");
                Emp emp2 = mapper.queryById(4);
                System.out.println("第二次查询结果：" + emp2);
                sqlSession1 = sqlSessionFactory.openSession();
                EmpMapper mapper1 = sqlSession1.getMapper(EmpMapper.class);
                Emp emp3 = mapper1.queryById(4);
                System.out.println("第三次查询结果：" + emp3);
            } finally {
                if (sqlSession != null) {
                    sqlSession.close();
                }
                if (sqlSession1 != null) {
                    sqlSession1.close();
                }
            }
        }
    }
```

上面代码中，总共执行了三次简单的查询操作，查询参数不变，然后对第一次查询出来的 Emp 对象的 username 进行了修改。但是并未提交，因此数据库数据还是不变的，然后使用同一个 SqlSession 对象再次获取查询数据，第三次使用新的 SqlSession 对象进行一次查询。控制台显示结果，如图 4-9 所示。

从控制台显示的结果来看，在并未修改数据库的情况下，第二次查询的结果和第一次查询结果并不一致，这是因为一级缓存的存在。在使用同一个 SqlSession 对象获取数据，第一次查询后会将结果缓存起来，后面再次通过相同的 SQL 语句且参数未发生变化的情况下，那么它就会从缓存中获取数据。而缓存中的对象就是 Emp，而我们在第一次查询后修改了 Emp 的 username 值，所以再次查询得到的就是图 4-9 所示的结果。

```
2020-06-07 16:26:42,404 [main] DEBUG [org.apache.ibatis.transaction.jdbc.JdbcTransaction] - Setting autocommit to false on JDBC
 Connection [com.mysql.cj.jdbc.ConnectionImpl@5a63f509]
2020-06-07 16:26:42,416 [main] DEBUG [com.bailiban.mapper.EmpMapper.queryById] - ==>  Preparing: select id eid, username,
 birthday, sex, address from tab_emp where id = ?
2020-06-07 16:26:42,576 [main] DEBUG [com.bailiban.mapper.EmpMapper.queryById] - ==> Parameters: 4(Integer)
2020-06-07 16:26:42,664 [main] DEBUG [com.bailiban.mapper.EmpMapper.queryById] - <==      Total: 1
第一次查询结果：Emp(id=4, username=张良, birthday=Sun Oct 25 00:00:00 CST 2009, sex=男, address=湖北武汉东湖高新, roles=null)
--------------分割线----------------
第二次查询结果：Emp(id=4, username=测试员工名, birthday=Sun Oct 25 00:00:00 CST 2009, sex=男, address=湖北武汉东湖高新, roles=null)
2020-06-07 16:26:42,666 [main] DEBUG [org.apache.ibatis.transaction.jdbc.JdbcTransaction] - Opening JDBC Connection
2020-06-07 16:26:42,678 [main] DEBUG [org.apache.ibatis.datasource.pooled.PooledDataSource] - Created connection 440737101.
2020-06-07 16:26:42,678 [main] DEBUG [org.apache.ibatis.transaction.jdbc.JdbcTransaction] - Setting autocommit to false on JDBC
 Connection [com.mysql.cj.jdbc.ConnectionImpl@1a451d4d]
2020-06-07 16:26:42,679 [main] DEBUG [com.bailiban.mapper.EmpMapper.queryById] - ==>  Preparing: select id eid, username,
 birthday, sex, address from tab_emp where id = ?
2020-06-07 16:26:42,679 [main] DEBUG [com.bailiban.mapper.EmpMapper.queryById] - ==> Parameters: 4(Integer)
2020-06-07 16:26:42,684 [main] DEBUG [com.bailiban.mapper.EmpMapper.queryById] - <==      Total: 1
第三次查询结果：Emp(id=4, username=张良, birthday=Sun Oct 25 00:00:00 CST 2009, sex=男, address=湖北武汉东湖高新, roles=null)
2020-06-07 16:26:42,685 [main] DEBUG [org.apache.ibatis.transaction.jdbc.JdbcTransaction] - Resetting autocommit to true on JDBC
 Connection [com.mysql.cj.jdbc.ConnectionImpl@5a63f509]
```

图 4-9

其实从控制台 SQL 输出语句中也可以看出，第二次查询并未向数据库发送 SQL 请求，而第三次查询使用了新的 SqlSession 对象来获取数据，因此发送了 SQL 请求，重新从数据库获取了数据。这也证明了一级缓存是和 SqlSession 绑定的，只存在于 SqlSession 的生命周期中。

注意：在 select 查询元素中加上 flushCache 属性，并配置为 true，就会在查询数据前清空当前的一级缓存。另外任何的 INSERT、UPDATE、DELETE 操作也都会清空一级缓存。

## 4.8.2 二级缓存

MyBatis 的二级缓存全局开关默认也是启用的，但是不同于一级缓存的是，还需要在要使用二级缓存的 Mapper 映射文件中添加<cache/>元素才能对当前 Mapper 使用二级缓存。如果想关闭二级缓存全局开关，可以在 mybatis-config.xml 中将 cacheEnabled 的值配置为 false，如下代码所示。这样即使在 Mapper 中添加了<cache/>元素也无法使用二级缓存。

```
<settings>
<!--是否开启二级缓存，默认值：true -->
    <setting name="cacheEnabled" value="false"/>
</settings>
```

下面使用代码来演示二级缓存的使用，在 EmpMapper.xml 中加入如下代码：

```
<?xml version="1.0" encoding="UTF-8"?>
<!DOCTYPE mapper PUBLIC "-//mybatis.org//DTD Mapper 3.0//EN"
"http://mybatis.org/dtd/mybatis-3-mapper.dtd">
<mapper namespace="com.bailiban.mapper.EmpMapper">
    <cache/>
<!--其他配置，此处省略-->
```

```
</mapper>
```

在测试类中进行测试，测试代码如下：

```
@Test
public void queryEmpById() {
    SqlSession sqlSession = null;
    SqlSession sqlSession1 = null;
    try {
        SqlSessionFactory sqlSessionFactory = SqlSessionFactoryWithXmlUtil.getSqlSessionFactory();
        // 生成 sqlSession
        sqlSession = sqlSessionFactory.openSession();
        // 使用 SqlSession 创建 mapper 接口的代理对象
        EmpMapper mapper = sqlSession.getMapper(EmpMapper.class);
        Emp emp1 = mapper.queryById(4);
        System.out.println("第一次查询结果：" + emp1);
        // 需要提交，如果是一级缓存，MyBatis 才会缓存对象到 SqlSessionFactory 层面
        sqlSession.commit();
        System.out.println("--------------分割线----------------");
        sqlSession1 = sqlSessionFactory.openSession();
        EmpMapper mapper1 = sqlSession1.getMapper(EmpMapper.class);
        Emp emp2 = mapper1.queryById(4);
        System.out.println("第二次查询结果：" + emp2);
    } finally {
        if (sqlSession != null) {
            sqlSession.close();
        }
        if (sqlSession1 != null) {
            sqlSession1.close();
        }
    }
}
```

控制台显示结果，如图 4-10 所示。

```
2020-06-07 17:38:15,870 [main] DEBUG [org.apache.ibatis.transaction.jdbc.JdbcTransaction] - Setting autocommit to false on JDBC
 Connection [com.mysql.cj.jdbc.ConnectionImpl@635eaaf1]
2020-06-07 17:38:15,880 [main] DEBUG [com.bailiban.mapper.EmpMapper.queryById] - ==>  Preparing: select id eid, username,
 birthday, sex, address from tab_emp where id = ?
2020-06-07 17:38:15,978 [main] DEBUG [com.bailiban.mapper.EmpMapper.queryById] - ==> Parameters: 4(Integer)
2020-06-07 17:38:16,040 [main] DEBUG [com.bailiban.mapper.EmpMapper.queryById] - <==      Total: 1
第一次查询结果：Emp(id=4, username=张良, birthday=Sun Oct 25 00:00:00 CST 2009, sex=男, address=湖北武汉东湖高新, roles=null)
--------------分割线----------------
2020-06-07 17:38:16,066 [main] DEBUG [com.bailiban.mapper.EmpMapper] - Cache Hit Ratio [com.bailiban.mapper.EmpMapper]: 0.5
第二次查询结果：Emp(id=4, username=张良, birthday=Sun Oct 25 00:00:00 CST 2009, sex=男, address=湖北武汉东湖高新, roles=null)
2020-06-07 17:38:16,066 [main] DEBUG [org.apache.ibatis.transaction.jdbc.JdbcTransaction] - Resetting autocommit to true on JDBC
 Connection [com.mysql.cj.jdbc.ConnectionImpl@635eaaf1]
```

图 4-10

开启二级缓存后，使用不同的 SqlSession 获取同一条记录，都只发送一次 SQL。因为此时 MyBatis 将其保存在 SqlSessionFactory 层面，可以提供给各个 SqlSession 使用。

注意：二级缓存中 MyBatis 会序列化和反序列化 POJO，因此我们的 POJO 类要实现

java.io.Serializable 接口，否则会抛出异常。

## 单元小结

- select 查询元素中 resultType 和 resultMap 之间只能同时使用一个且必须使用。
- MyBatis 映射接口方法中可以使用@Param 注解进行参数的传递。
- MyBatis 级联有三种：鉴别器(discriminator)、一对一(association)、一对多(collection)。
- MyBatis 中一级缓存是默认开启的，且不可关闭。
- 一级缓存是 SqlSession 层面的，二级缓存是 SqlSessionFactory 层面的。

## 单元自测

1. 在使用 MyBatis 的时候，除了可以使用@Param 注解来实现多参数入参，还可以用(　　)传递多个参数值。

   A. 用 Map 对象

   B. 用 List 对象

   C. 用数组的方式

   D. 用 Set 集合的方式

2. MyBatis 是通过(　　)进行分页的。

   A. RowBounds　　　　B. limit

   C. ResultSet　　　　D. PageHelper

3. 当使用 MySql 的主键自增方式时，使用 selectKey 进行主键回填，order 值为(　　)。

   A. AFTER　　B. BEFORE　　C. 可以省略　　D. AFTER 或 BEFORE

4. 以下关于 MyBatis，正确的说法是(　　)。

   A. 不支持延迟加载

   B. 仅支持 association 关联对象的延迟加载

   C. 仅支持 collection 关联集合对象的延迟加载

   D. 仅支持 association 关联对象和 collection 关联集合对象的延迟加载

## 上机实战

### 上机目标

- 掌握 MyBatis 中主键回填的方法。
- 掌握级联的配置。

### 上机练习

◆ 第一阶段◆

#### 练习 1：插入 Card 工号信息及对应的 Emp 员工信息

**【问题描述】**

新进员工时，先有员工个人信息，随后办理入职后才会有相应的工号信息。下面模拟这种情况进行添加操作。

**【问题分析】**

(1) 先向 tab_emp 表中插入 Emp 员工信息，并获取其主键(这里采用主键回填，且主键是自增的)。

(2) 再向 tab_card 表中插入 Card 工号信息。

**【参考步骤】**

(1) 新建一个工程：参考 mybatis-demo3 工程。

(2) 在 EmpMapper.xml 和 CardMapper.xml 中加入 insert 插入语句。代码如下所示：

EmpMapper.xml 中加入：

```
<insert id="insertEmp">
    <selectKey keyProperty="id" resultType="int" order="AFTER">
        select LAST_INSERT_ID()
    </selectKey>
    insert into tab_emp(username, birthday, sex, address)
    VALUES (#{username}, #{birthday}, #{sex}, #{address})
</insert>
```

CardMapper.xml 中加入：

```
<insert id="insertCard">
    insert into tab_card(card, create_time, eid)
    values (#{card}, #{createTime}, #{emp.id})
</insert>
```

(3) 在 EmpMapper 和 CardMapper 接口中加入相应的映射接口方法。代码如下所示：

EmpMapper 中加入：

```
/**
 * 插入 Emp
 *
 * @param emp emp 对象
 * @return 受影响行数
 */
int insertEmp(Emp emp);
```

CardMapper 中加入：

```
/**
 * 插入 Card
 *
 * @param card card 对象
 * @return 受影响行数
 */
int insertCard(Card card);
```

(4) 在测试代码进行测试，代码如下所示：

```
@Test
public void insertCard() {
    SqlSession sqlSession = null;
    try {
        SqlSessionFactory sqlSessionFactory = SqlSessionFactoryWithXmlUtil.getSqlSessionFactory();
        // 生成 sqlSession
        sqlSession = sqlSessionFactory.openSession();
        // 使用 SqlSession 创建 mapper 接口的代理对象
        CardMapper cardMapper = sqlSession.getMapper(CardMapper.class);
        EmpMapper empMapper = sqlSession.getMapper(EmpMapper.class);
        Emp emp = new Emp();
        emp.setUsername("佳琪");
        emp.setAddress("浙江杭州滨江新园区");
        LocalDate localDate = LocalDate.of(1987, 12, 12);
        emp.setBirthday(Date.from(localDate.atStartOfDay(ZoneId.systemDefault()).toInstant()));
        emp.setSex("男");
        // 插入 emp
        empMapper.insertEmp(emp);
        Card card = new Card();
        card.setCard("c10021");
        card.setCreateTime(new Date());
        card.setEmp(emp);
        // 插入 card
        cardMapper.insertCard(card);
        sqlSession.commit();
```

```
        } finally {
            if (sqlSession != null) {
                sqlSession.close();
            }
        }
    }
```

(5) 查看控制台的输出。

## ◆第二阶段◆

### 练习 2：查询员工信息。级联查询出其工号信息，同时懒加载其对应的角色和角色处于可用状态下的权限信息

**【问题描述】**

这是一个对级联和懒加载的综合应用，可以参照 mybatis-demo3 中的代码进行解答。

**【问题分析】**

(1) 考察员工对工号一对一关系。

(2) 考察懒加载和立即加载的配置。

(3) 考察员工对角色一对多，角色对权限一对多的关系。

(4) 考察 discriminator 鉴别器的使用。

**【参考步骤】**

(1) 使用 association 标签完成员工和工号的一对一级联查询。

(2) 在级联查询工号信息时设置 fetchType="eager"。

(3) 使用 collection 标签完成员工对角色的一对多级联查询。

(4) 使用 discriminator 标签完成角色可用状态下选择映射权限信息结果集。

# 动态SQL

## 课程目标

❖ 掌握 MyBatis 动态 SQL 的基本使用

❖ 掌握 MyBatis 动态 SQL 的基本元素的用法

❖ 掌握 MyBatis 动态 SQL 的条件判断方法

**简介**

动态 SQL 是 MyBatis 的强大特性之一。使用过 JDBC 或其他类似的框架的用户，应该能理解根据不同条件拼接 SQL 语句的痛苦，例如拼接时要确保不能忘记添加必要的空格，还要注意去掉列表最后一个列名的逗号。利用 MyBatis 的动态 SQL，可以彻底摆脱这种痛苦。

在 MyBatis 3 之前的版本中，需要花时间了解大量的元素。借助功能强大的基于 OGNL 的表达式，MyBatis 3 替换了之前的大部分元素，大大精简了元素种类，现在要学习的元素种类比原来的一半还少。以下是 MyBatis 的动态 SQL 在 XML 中支持的几种标签。

- if
- choose、when、otherwise
- trim、where、set
- foreach
- bind

下面就逐一对这些标签进行讲解。

## 5.1 if 标签

if 标签是最常用的判断语句，相当于 Java 中的 if 语句，使用起来非常简单。

假如有一个场景：根据 tab_emp 表中的员工姓名第一个字和地址模糊查询这两个条件去查询员工信息，但地址是一个选填条件，不填写时，就不要用它作为条件查询。

假如按照以前的查询方式，代码可能写成下面这样：

```
<select id="queryEmpsByFirstNameAndAddress" resultMap="EmpMap">
    select id eid,
            username,
            birthday,
            sex,
            address
    from tab_emp
    where username like concat(#{firstName}, '%')
      and address like concat('%',#{address}, '%')
</select>
```

当同时输入 firstName 和 address 两个条件时，能查询出正确的结果，但当只提供 firstname 的参数值时，address 设置为 null，就无法查询到正确结果。因为 address=null 也作为了查询条件。此时就可以使用 if 标签来解决此类问题。代码如下所示：

```
<select id="queryEmpsByFirstNameAndAddress" resultMap="EmpMap">
    select id eid,
    username,
    birthday,
    sex,
    address
    from tab_emp
    where username like concat(#{firstName}, '%')
    <if test="address!=null and address!='' ">
        and address like concat('%',#{address}, '%')
    </if>
</select>
```

if 标签有一个必填的属性 test，test 的属性值是一个符合 OGNL 要求的判断表达式，表达式的结果为 true 或 false，除此之外所有的非 0 值都为 true，只有 0 为 false。

上面代码中判断条件 address 不为空并且 address 不为空字符串，满足这些条件后，才会将 address 作为查询条件。这才符合查询要求。

## 5.2 choose、when、otherwise 标签

在 5.1 中我们使用 if 标签的条件判断，但该标签无法实现经常用到的其他条件的判断逻辑，这就需要用到 choose、when、otherwise 这三个标签。我们看下面一段代码：

```
<select id="queryEmpsByFirstNameOrAddress" resultMap="EmpMap">
    select id eid,
    username,
    birthday,
    sex,
    address
    from tab_emp
    where 1=1
    <choose>
        <when test="firstName!=null and firstName!=''">
            and username like concat(#{firstName}, '%')
        </when>
        <when test="address!=null and address!='' ">
            and address like concat('%',#{address}, '%')
        </when>
```

```
            <otherwise>
                and sex='男'
            </otherwise>
        </choose>
    </select>
```

这段代码表达的意思是：查询员工信息，先按照 firstName 作为条件查询，当 firstName 为空或空字符串时才会使用 address 作为条件查询，如果 address 也为空或空字符串，那么就按照 sex='男'作为条件查询。

## 5.3 where、trim、set 标签

在 5.2 小节中的动态 SQL 语句中加入了一个 1=1 条件，假如没有这个条件就可能会变成这样一条错误的语句：

```
select id eid,
username,
birthday,
sex,
address
from tab_emp
where and username like concat(#{firstName}, '%')
```

这条 SQL 很明显会报出关于 SQL 的语法异常。虽然加入 1=1 能解决这个异常，但是不免显得有些奇怪，此时可以使用 where 标签来处理，代码如下所示。

```
<select id="queryEmpsByFirstNameOrAddress" resultMap="EmpMap">
    select id eid,
    username,
    birthday,
    sex,
    address
    from tab_emp
    <where>
        <choose>
            <when test="firstName!=null and firstName!=''">
                and username like concat(#{firstName}, '%')
            </when>
            <when test="address!=null and address!=' '">
                and address like concat('%',#{address}, '%')
            </when>
            <otherwise>
                and sex='男'
            </otherwise>
```

```
            </choose>
        </where>
    </select>
```

当 where 标签里面的条件成立时，才会加入 where 关键字组装到 SQL 里面且会将紧跟 where 后面的第一个 and 自动去掉，否则就不会加入。

其实上面这种去掉多余的 SQL 关键字或符号，比如常见的 and、or 和逗号，使用 trim 也可以完成。代码如下所示。

```
<select id="queryEmpsByFirstNameOrAddress" resultMap="EmpMap">
    select id eid,
    username,
    birthday,
    sex,
    address
    from tab_emp
    <trim prefix="where" prefixOverrides="and">
        <choose>
            <when test="firstName!=null and firstName!=''">
                and username like concat(#{firstName}, '%')
            </when>
            <when test="address!=null and address!=' '">
                and address like concat('%',#{address}, '%')
            </when>
            <otherwise>
                and sex='男'
            </otherwise>
        </choose>
    </trim>
</select>
```

关于 trim 标签属性解释如下：

- prefix：当 trim 元素内包含内容时，会给内容增加 prefix 指定的前缀。
- prefixOverrides：当 trim 元素内包含内容时，会把内容中匹配的前缀字符串去掉。
- suffix：当 trim 元素内包含内容时，会给内容增加 suffix 指定的后缀。
- suffixOverrides：当 trim 元素内包含内容时，会把内容中匹配的后缀字符串去掉。

上面代码中使用 trim 标签的写法与 where 标签是等效的。

在另外一种场景中我们也可以用到 trim 标签，那就是更新操作，很多时候在执行更新时，只是想更新某些字段，而非全部字段。代码如下所示。

```
<update id="updateEmp">
    update tab_emp
    <trim prefix="set" suffixOverrides=",">
        <if test="username!=null and username!=''">
```

```
            username=#{username},
        </if>
        <if test="birthday!=null and birthday!=''">
            birthday=#{birthday},
        </if>
        <if test="sex!=null and sex!=''">
            sex=#{sex},
        </if>
        <if test="address!=null and address!=''">
            address=#{address}
        </if>
    </trim>
    where id = #{id}
</update>
```

上面代码除了使用 trim 标签外，还可以使用另外一种标签：set 标签。代码如下所示。

```
<update id="updateEmp">
    update tab_emp
    <set>
        <if test="username!=null and username!=''">
            username=#{username},
        </if>
        <if test="birthday!=null and birthday!=''">
            birthday=#{birthday},
        </if>
        <if test="sex!=null and sex!=''">
            sex=#{sex},
        </if>
        <if test="address!=null and address!=''">
            address=#{address}
        </if>
    </set>
    where id = #{id}
</update>
```

set 标签的作用就是如果该标签包含的元素中有返回值，就插入一个 set，如果 set 后面的字符串是以逗号结尾的，就将这个逗号删除。

## 5.4 foreach 标签

动态 SQL 的另一个常见使用场景是对集合进行遍历(尤其是在构建 IN 条件语句的时候)。此时就可以使用 foreach 标签来处理，代码如下所示。

```
<select id="findRolesByIds" resultMap="RoleMap">
    select tr.id rid,
    tr.role_name,
    tr.role_desp,
    tr.enable
    from tab_role tr
    where id in
    <foreach item="id" index="index" collection="ids" open="(" separator="," close=")">
        #{id}
    </foreach>
</select>
```

foreach 标签包含的属性解释如下：

- collection：必选，值为要迭代循环的属性名。
- item：集合中元素迭代时的别名。
- index：集合中元素迭代时的索引。
- open：整个循环内容开头的字符串。
- close：整个循环内容结尾的字符串。
- separator：每次循环的分隔符。

在测试类中进行测试，测试代码如下：

```
@Test
public void findRolesByIds() {
    SqlSession sqlSession = null;
    try {
        SqlSessionFactory sqlSessionFactory = SqlSessionFactoryWithXmlUtil.getSqlSessionFactory();
        // 生成 sqlSession
        sqlSession = sqlSessionFactory.openSession();
        //  使用 SqlSession 创建 mapper 接口的代理对象
        RoleMapper mapper = sqlSession.getMapper(RoleMapper.class);
        Integer[] ids = {1, 2, 3, 4};
        List<Role> roles = mapper.findRolesByIds(ids);
        for (Role role : roles) {
            System.out.println(role);
        }
    } finally {
        if (sqlSession != null) {
            sqlSession.close();
        }
    }
}
```

控制台的输出结果如图 5-1 所示。

```
2020-06-10 22:18:53,231 [main] DEBUG [org.apache.ibatis.transaction.jdbc.JdbcTransaction] - Setting autocommit to false on JDBC
 Connection [com.mysql.cj.jdbc.ConnectionImpl@402bba4f]
2020-06-10 22:18:53,242 [main] DEBUG [com.bailiban.mapper.RoleMapper.findRolesByIds] - ==>  Preparing: select tr.id rid,
 tr.role_name, tr.role_desp, tr.enable from tab_role tr where id in ( ? , ? , ? , ? )
2020-06-10 22:18:53,373 [main] DEBUG [com.bailiban.mapper.RoleMapper.findRolesByIds] - ==> Parameters: 1(Integer), 2(Integer),
 3(Integer), 4(Integer)
2020-06-10 22:18:53,565 [main] DEBUG [com.bailiban.mapper.RoleMapper.findRolesByIds] - <==      Total: 4
Role(id=1, roleName=董事长, roleDesp=企业老大, enable=1, emps=null, permissions=null)
Role(id=2, roleName=总经理, roleDesp=管理整个公司, enable=1, emps=null, permissions=null)
Role(id=3, roleName=副总经理, roleDesp=分管其中一个片区, enable=1, emps=null, permissions=null)
Role(id=4, roleName=人力hr, roleDesp=负责招聘员工组织管理, enable=0, emps=null, permissions=null)
2020-06-10 22:18:53,567 [main] DEBUG [org.apache.ibatis.transaction.jdbc.JdbcTransaction] - Resetting autocommit to true on JDBC
 Connection [com.mysql.cj.jdbc.ConnectionImpl@402bba4f]
```

图 5-1

从控制台输出的 SQL 语句也可以看出，使用 foreach 这种方式还可以防止 SQL 注入。

## 5.5 bind 标签

bind 元素允许在 OGNL 表达式以外创建一个变量，并将其绑定到当前的上下文。例如，在前面进行地址的模糊查询时，用到一个关键字 concat 函数，若是换成 Oracle 数据库则会报错，因为 concat 函数在 Oracle 数据库中有且仅有两个参数，在 MySQL 数据库中可以有多个参数。使用 bind 就可以解决这种情况。代码如下所示。

```
<select id="queryEmpsByFirstNameAndAddress" resultMap="EmpMap">
    <bind name="username" value="firstName+'%'"/>
    select id eid,
    username,
    birthday,
    sex,
    address
    from tab_emp
    where username like #{username}
    <if test="address!=null and address!='' ">
        <bind name="address" value="'%'+address+'%'"/>
        and address like #{address}
    </if>
</select>
```

bind 标签的两个属性都是必填项，name 为绑定到上下文的变量名，value 为 OGNL 表达式。

在测试类中进行测试，测试代码如下：

```
@Test
public void queryEmpsByFirstNameAndAddress() {
    SqlSession sqlSession = null;
    try {
```

```
            SqlSessionFactory sqlSessionFactory = SqlSessionFactoryWithXmlUtil.getSqlSessionFactory();
            // 生成 sqlSession
            sqlSession = sqlSessionFactory.openSession();
            // 使用 SqlSession 创建 mapper 接口的代理对象
            EmpMapper mapper = sqlSession.getMapper(EmpMapper.class);
            List<Emp> emps = mapper.queryEmpsByFirstNameAndAddress("张", "武汉");
            for (Emp emp : emps) {
                System.out.println(emp);
            }
        } finally {
            if (sqlSession != null) {
                sqlSession.close();
            }
        }
    }
```

控制台的输出结果如图 5-2 所示。

```
2020-06-10 23:08:19,222 [main] DEBUG [org.apache.ibatis.transaction.jdbc.JdbcTransaction] - Setting autocommit to false on JDBC
 Connection [com.mysql.cj.jdbc.ConnectionImpl@5fdcaa40]
2020-06-10 23:08:19,233 [main] DEBUG [com.bailiban.mapper.EmpMapper.queryEmpsByFirstNameAndAddress] - ==>  Preparing: select id
 eid, username, birthday, sex, address from tab_emp where username like ? and address like ?
2020-06-10 23:08:19,350 [main] DEBUG [com.bailiban.mapper.EmpMapper.queryEmpsByFirstNameAndAddress] - ==> Parameters: 张%(String),
 %武汉%(String)
2020-06-10 23:08:19,440 [main] DEBUG [com.bailiban.mapper.EmpMapper.queryEmpsByFirstNameAndAddress] - <==      Total: 2
Emp(id=4, username=张良, birthday=Sun Oct 25 00:00:00 CST 2009, sex=男, address=湖北武汉东湖高新, roles=null)
Emp(id=7, username=张芬, birthday=Fri Jan 01 00:00:00 CST 1988, sex=女, address=湖北武汉东西湖区, roles=null)
2020-06-10 23:08:19,529 [main] DEBUG [org.apache.ibatis.transaction.jdbc.JdbcTransaction] - Resetting autocommit to true on JDBC
 Connection [com.mysql.cj.jdbc.ConnectionImpl@5fdcaa40]
```

图 5-2

- if 标签中 test 为必填属性，property!=''或 property==''仅适用于 String 类型的字段。
- where 和 set 标签的功能都可以用 trim 标签来实现。
- foreach 标签中的 collection 为必填属性，其属性值的情况有多种。

1. 下列标签中(　　)不是 MyBatis 动态 SQL 语句中的标签。

   A. id　　　　B. if

   C. otherwise　　　　D. where

2. 关于 MyBatis 的 trim 元素，下列说法正确的有(　　)。

   A. prefix 属性的作用是在 trim 元素包含的内容前加上某些前缀

B. suffix 属性的作用是在 trim 元素包含的内容前加上某些后缀

C. prefixOverrides 的作用是把 trim 元素包含内容的首部某些内容覆盖，即忽略不使用该字段

D. suffixOverrides 的作用是把 trim 元素包含内容的尾部某些内容覆盖，即忽略不使用该字段

3. 关于 MyBatis 的下列说法中，正确的选项有(　　)。

A. choose 标签类似于 Java 的 switch 语句

B. 利用 trim 可以代替 where/set 标签的功能

C. foreach 的 collection 属性可以取的值有 list、array

D. bind 标签可以使用 OGNL 表达式创建一个变量并将其绑定到上下文中

4. 关于 MyBatis 的 foreach 元素，下列(　　)不是它的属性。

A. item　　B. index

C. end　　D. collection

## 上机实战

### 上机目标

- 了解 if 标签中关于字符串的判断。
- 学会使用 foreach 标签实现批量插入。

### 上机练习

◆ 第一阶段◆

#### 练习 1：用 test 属性判断字符串

【问题分析】

(1) test 用于条件判断，大多时候用来判断空和非空，有时候也会用来判断数字、字符串等。

(2) 假设一个场景，根据性别 sex 查询 tab_emp 表，且当性别为男性的时候还加上地址模糊查询条件，否则就不加。

(3) 对于字符串的判断可以加上 toString()的方法进行比较。

**【参考步骤】**

(1) 新建工程参照 mybatis-demo3。

(2) 在 EmpMapper.xml 中加入如下代码。

```
<select id="queryEmpsBySex" resultMap="EmpMap">
    select id eid,
    username,
    birthday,
    sex,
    address
    from tab_emp
    <where>
        <if test="sex!=null and sex!=''">
            sex=#{sex}
        </if>
        <if test="sex=='男'.toString()">
            <bind name="address" value="'%'+address+'%'"/>
            and address like #{address}
        </if>
    </where>
</select>
```

(3) 在 EmpMapper 接口中加入如下代码。

```
/**
 * 根据性别查询，性别为男就再加上地址条件
 *
 * @param sex         性别
 * @param address  地址
 * @return emps
 */
List<Emp> queryEmpsBySex(@Param("sex") String sex, @Param("address") String address);
```

**◆ 第二阶段◆**

## 练习 2：使用 foreach 标签实现批量插入

**【问题描述】**

批量插入数据也是经常会用到的一种场景，在 MySQL 数据库中批量插入的语法如下所示。

```
INSERT INTO tab_emp ( username, birthday, sex, address )
VALUES ( '姓名 1', '1996-12-12', '男', '地址 1' ),
 ( '姓名 2', '1996-12-12', '男', '地址 2' ),
( '姓名 3', '1996-12-12', '男', '地址 3' ),
```

```
( '姓名 4', '1996-12-12', '男', '地址 4' )
......
```

从上面的SQL语句中可以看出插入的数据部分是一个值的循环，因此可以通过foreach实现批量插入。

**【问题分析】**

(1) 使用foreach标签，需要一个对象的集合list。

(2) foreach标签中除了collection必填外，只要一个item和separator属性即可。

**【参考步骤】**

根据问题分析参考练习一完成。

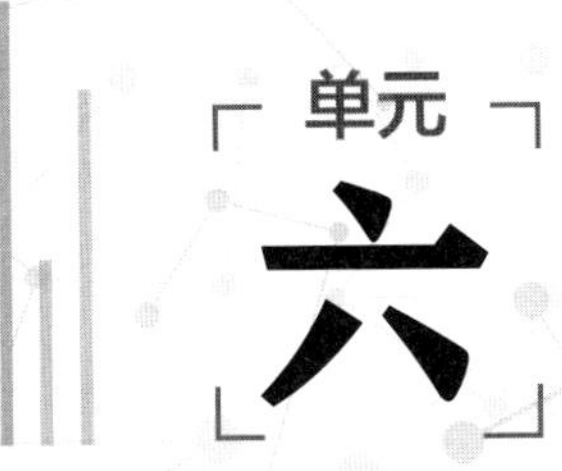


# Bean工厂与应用上下文

## 课程目标

- ❖ 理解控制反转与依赖注入的原理
- ❖ 掌握 Spring 核心 Bean 工厂
- ❖ 掌握 Bean 装配中注解的使用
- ❖ 掌握 Bean 的生命周期

**简介**

编写程序要尽量避免应用代码对容器的依赖，但是通常情况下只有非常简单的对象能够独立工作，大多数业务对象都是有依赖关系的，要么与其他受容器管理的业务对象、与数据访问对象存在依赖关系，要么需要获取各种资源。通常情况下，对象是使用容器提供的查找能力来定位它所依赖的受管对象或资源，要解决对象间的依赖问题则必须在代码中使用容器提供的查找对象的接口，似乎受管对象必须得依赖容器。然而现实中，可以使用反转控制(Inverse of Control，IoC)来解决依赖而又使对象不依赖于容器。

# 6.1 Spring IoC 概述

在框架中，控制反转并不是一个什么新东西，它是一个很普通但非常重要的概念，设计模式中的模板方法设计模式就是控制反转的一个示例，其更为典型的形式是所谓的好莱坞原则：“不要打电话给我们，我们会打电话给你”；更为简单的理解是 IoC 并不用去创建对象，对象都由工厂来提供，只要告诉工厂需求，即使所要的东西是由多个部件组装而成的，工厂也能提供；IoC 有依赖查找和依赖注入两种最主要的实现方式。

## 6.1.1 依赖查找

依赖查找(Dependency Lookup)指容器中的受控对象通过容器的 API 查找自己所依赖的资源和协作对象。这种方式虽然降低了对象间的依赖，但是同时也使用了容器的 API，使得无法在容器外使用和测试对象。依赖查找是一种更加传统的 IoC 实现方式。

## 6.1.2 依赖注入

控制反转有第二种实现策略——依赖注入，相比依赖查找，它是更可取的方式。这种方式让容器去负责查找依赖，并且被 IoC 容器管理的对象在初始化时通过暴露出来的 JavaBean 的 setter 方法或是带参数的构造方法进行组装，解决对象间的依赖，并且不需要依赖特定容器的 API 或接口。

依赖注入的基本原则是：应用对象本身应该负责查找资源或其他依赖的对象，配置对象的依赖关系的工作应该由 IoC 容器完成，“查找依赖”的代码就从业务代码中剥离出来，交给容器负责。这样，使用依赖注入时，IoC 容器代替我们完成了依赖查找的工作。如果需要不同的方式获得资源或是设置不同的依赖对象，只需要对容器重新配置就可以了，不必修改业务代码。

使用依赖注入好处是：

(1) 查找和装配依赖关系的代码完全与业务代码无关。

(2) 不需要依赖容器的特定 API，可以很简单地在容器外使用业务对象。

(3) 不需要使用特殊的接口，全都是普通的 Java 对象，大大降低了业务代码对 IoC 容器的依赖，并且大多数情况下业务对象可以做到完全不依赖容器。

(4) 对于业务对象所依赖的类似于 String、int 这样的简单数据类型，也可以采用同样的方式处理。

通常情况，我们可以使用两种注入方式来完成对象的依赖：设值注入和构造注入。下面我们分别来进行讲解。

### 1. 在 Idea 中使用 Spring

在讲解设值注入和构造注入前，先演示如何在 Idea 工具中开发 Spring 应用。

(1) 先在 Idea 中创建一个基本的 Maven 项目，命名为 spring-demo，可以勾选模板 maven-archetype-quickstart，具体步骤请参照本书单元二中的相关内容。

(2) 创建好 Maven 项目后，在 pom 文件中添加 Spring 相关依赖坐标，代码如下所示。

```
<dependency>
    <groupId>org.springframework</groupId>
    <artifactId>spring-context</artifactId>
    <version>5.2.7.RELEASE</version>
</dependency>
<dependency>
    <groupId>junit</groupId>
    <artifactId>junit</artifactId>
    <version>4.11</version>
    <scope>test</scope>
</dependency>
```

此时我们的项目准备工作就完成了。

### 2. 设值注入

设值注入是指 IoC 容器通过成员变量的 setter 方法来注入被依赖对象。这种注入方式简单、直观，因此被大量使用。

下面示例代码展示了设值注入的使用。

```
public class UserServiceImpl implements UserService {
    private UserDao userDao;

    /**
     * 设值注入所需要的 setter 方法
     * @param userDao
     */
    public void setUserDao(UserDao userDao) {
        this.userDao = userDao;
    }

    @Override
    public void login() {
        userDao.login();
    }
}
```

上面程序中，在业务实现类中，login 方法需要调用 DAO 持久层接口方法，因此可以通过 setter 方法注入所需要的 UserDao。而 UserServiceImpl 实现类并不知道要调用的 userDao 实例在哪里，也不知道 userDao 实例是如何实现的，它只负责调用 UserDao 对象的方法，这个 userDao 实例将由 Spring 容器负责注入。

### 3. 构造注入

构造注入就是通过 Bean 类的构造方法，将 Bean 所依赖的对象注入。其本质就是驱动 Spring 在底层以反射方式执行带参数的构造器，当执行带参数的构造器时，就可以利用构造器参数对成员变量执行初始化。

下面对设值注入的代码进行修改，如下所示。

```
public class UserServiceImpl implements UserService {
    private UserDao userDao;

    /**
     * 构造注入所需的带参数的构造器
     *
     * @param userDao
     */
    public UserServiceImpl(UserDao userDao) {
        this.userDao = userDao;
    }

    @Override
    public void login() {
```

```
            userDao.login();
        }
    }
```

上面程序中没有提供设置 userDao 成员变量的 setter 方法，仅仅提供了一个带 UserDao 参数的构造器，Spring 将通过该构造器为 UserServiceImpl 注入所依赖的 Bean 实例。

## 6.2 Spring IoC 容器

Spring 框架对设值方法依赖注入和构造方法注入都提供了良好的支持。其中，两个最基本、最重要的包是 org.springframework.beans 和 org.springframework.context，在这两个包中，为了实现无侵入的框架，大量使用了反射机制，通过动态调用避免了硬编码，为 Spring 的反转控制特性提供了基础。

在这两个包中，最重要的类是 BeanFactory 和 ApplicationContext。BeanFactory 提供了一种先进的配置机制来管理任何种类的业务对象，这些业务对象在 Spring 中被称为 bean。ApplicationContext 建立在 BeanFactory 之上，并增加了其他的功能。

简而言之，BeanFactory 提供了配置框架和基本的功能，而 ApplicationContext 为它增加了更强的功能，这些功能中的一些或许更加接近 J2EE 并且围绕企业级应用。一般来说，ApplicationContext 是 BeanFactory 的完全超集，任何 BeanFactory 功能和行为的描述也同样被认为适用于 ApplicationContext。

## 6.3 ApplicationContext 简介

一般情况，都不会使用 BeanFactory 实例作为 Spring 容器，而是使用 ApplicationContext 实例作为容器，因此 Spring 容器也被称为 Spring 上下文。相对于 BeanFactory 来说，ApplicationContext 除了提供 BeanFactory 所有的功能外，还有一些其他的包括国际化支持、资源访问等额外功能。

ApplicationContext 是 Spring 容器最常用的接口，该接口的常用实现类有如下三个。

- ClassPathXmlApplicationContext：从类加载路径下搜索配置文件，并根据配置文件创建 Spring 容器。
- FileSystemXmlApplicationContext：从文件系统的相对路径或绝对路径下搜索配置文件，并根据配置文件来创建 Spring 容器。

- AnnotationConfigApplicationContext：基于 Java 的配置类(包括各种注解)加载 Spring 的应用上下文。避免使用 XML 进行配置。相比 XML 配置，更加便捷。

对于 Java 项目而言，类加载路径总是稳定的，因此通常总是使用 ClassPathXmlApplicationContext 创建 Spring 容器。代码如下所示。

```
public class AppTest {

    @Test
    public void testSetter() {
        ApplicationContext context = new ClassPathXmlApplicationContext("beans.xml");
        UserService userService = context.getBean("userServiceImpl", UserService.class);
        userService.login();
    }
}
```

接下来介绍如何通过 XML 组装 Bean。

在上面的代码中，我们使用到了 beans.xml 配置文件、UserService 接口、UserDao 接口和对应的实现类。其代码分别如下所示：

### 1. UserService 接口

```
public interface UserService {
    /**
     * 登录
     */
    void login();
}
```

### 2. UserDao 接口

```
public interface UserDao {
    /**
     * 登录
     */
    void login();
}
```

### 3. UserDaoImpl 实现类

```
public class UserDaoImpl implements UserDao {
    @Override
    public void login() {
        System.out.println("登录成功！");
    }
}
```

#### 4. UserServiceImpl 实现类

```
public class UserServiceImpl implements UserService {
    private UserDao userDao;

    /**
     * 设值注入所需的 setter 方法
     *
     * @param userDao
     */
    public void setUserDao(UserDao userDao) {
        this.userDao = userDao;
    }

    @Override
    public void login() {
        userDao.login();
    }

}
```

#### 5. beans.xml

```
<?xml version="1.0" encoding="UTF-8"?>
<beans xmlns="http://www.springframework.org/schema/beans"
       xmlns:xsi="http://www.w3.org/2001/XMLSchema-instance"
       xsi:schemaLocation="http://www.springframework.org/schema/beans
                           http://www.springframework.org/schema/beans/spring-beans.xsd">
<bean id="userServiceImpl" class="com.bailiban.service.impl.UserServiceImpl">
        <property name="userDao" ref="userDaoImpl"></property>
    </bean>
    <bean id="userDaoImpl" class="com.bailiban.dao.impl.UserDaoImpl"></bean>
</beans>
```

上面代码中，beans.xml 是我们接下来要重点讲解的内容。

## 6.4 Bean 的配置

Spring 采用了 XML 配置文件来指定实例之间的依赖关系，从 Spring 2.0 开始，Spring 推荐采用 XML Schema 来定义配置文件的语义约束。

在 Idea 中创建 Spring 的基础配置文件非常简单，如图 6-1 所示。

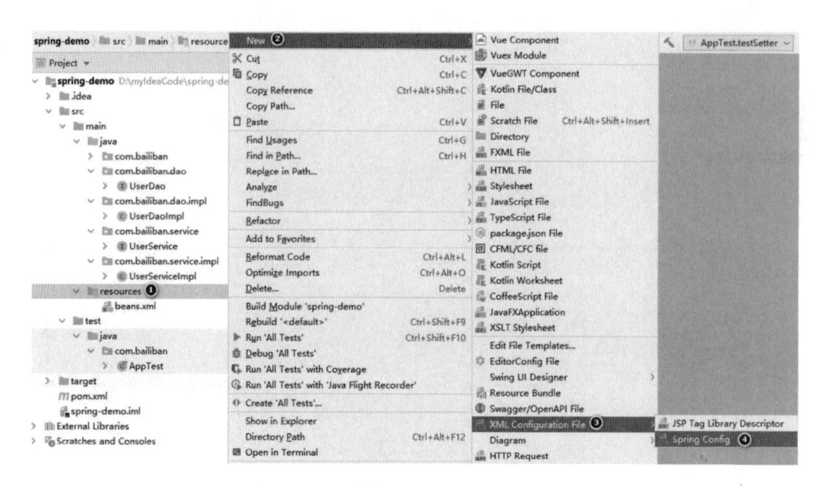


图 6-1

这样就可以创建好一个简单的配置，代码如下：

```
<?xml version="1.0" encoding="UTF-8"?>
<beans xmlns="http://www.springframework.org/schema/beans"
        xmlns:xsi="http://www.w3.org/2001/XMLSchema-instance"
        xsi:schemaLocation="http://www.springframework.org/schema/beans
                            http://www.springframework.org/schema/beans/spring-beans.xsd">

</beans>
```

在上述代码中引入了一个 beans 的定义，它是一个根元素，而 XSD 文件也被引入了，这样它所定义的元素将可以定义对应的 Spring Bean。

## 6.4.1 Bean 的标识

在 6.3 节中我们创建了一个完整的 beans.xml 配置文件，代码如下：

```
<?xml version="1.0" encoding="UTF-8"?>
<beans xmlns="http://www.springframework.org/schema/beans"
        xmlns:xsi="http://www.w3.org/2001/XMLSchema-instance"
        xsi:schemaLocation="http://www.springframework.org/schema/beans
                            http://www.springframework.org/schema/beans/spring-beans.xsd">
<bean id="userServiceImpl" class="com.bailiban.service.impl.UserServiceImpl">
        <property name="userDao" ref="userDaoImpl"></property>
    </bean>
    <bean id="userDaoImpl" class="com.bailiban.dao.impl.UserDaoImpl"></bean>
</beans>
```

在配置文件中，Spring 配置 Bean 实例通过 id 作为其唯一标识，程序通过 id 属性值访问该 Bean 实例。

当然也可以使用 name 属性来指定 Bean 的 id，代码如下：

```
<?xml version="1.0" encoding="UTF-8"?>
<beans xmlns="http://www.springframework.org/schema/beans"
        xmlns:xsi="http://www.w3.org/2001/XMLSchema-instance"
        xsi:schemaLocation="http://www.springframework.org/schema/beans
                            http://www.springframework.org/schema/beans/spring-beans.xsd">
    <bean name="userServiceImpl" class="com.bailiban.service.impl.UserServiceImpl">
        <property name="userDao" ref="userDaoImpl"></property>
    </bean>
    <bean name="userDaoImpl" class="com.bailiban.dao.impl.UserDaoImpl"></bean>
</beans>
```

使用 id 和 name 的区别在于：

id 属性允许指定一个 Bean 的 id，并且它在 XML DTD 中作为一个真正的 XML 元素的 ID 属性被标记，所以 XML 解析器能够作一些额外的校验，比如唯一性检查等。但是，XML 规范中对于 XML 的 ID 中的合法字符有比较严格的限定。如果要使用不符合 XML ID 限定的字符，则可以使用 name 属性。

## 6.4.2 Bean 的类

在上面的配置文件中，每一个 Bean 还有另外一个属性 class，它指定该 Bean 的实现类，此处不可再用接口，必须使用实现类，Spring 容器会使用 XML 解析器读取该属性值，并利用反射创建该实现类的实例。

## 6.4.3 Bean 的作用域

当通过 Spring 容器创建一个 Bean 实例时，不仅可以完成 Bean 实例的实例化，还可以为 Bean 指定特定的作用域。Spring 支持如下 6 种作用域。

- singleton：单例模式，在整个 Spring IoC 容器中，singleton 作用域的 Bean 将只生成一个实例。
- prototype：每次通过容器的 getBean()方法获取 prototype 作用域的 Bean 时，都将产生一个新的 Bean 实例。
- request：对于一次 HTTP 请求，request 作用域的 Bean 将只生成一个实例，这意味着，在同一次 HTTP 请求内，程序每次请求该 Bean，得到的总是同一个实例。只有在 Web 应用中使用 Spring 时，该作用域才真正生效。
- session：对于一次 HTTP 会话，session 作用域的 Bean 将只生成一个实例，这意味着，在同一次 HTTP 会话内，程序每次请求该 Bean，得到的总是同一个实例。只有在 Web 应用中使用 Spring 时，该作用域才真正生效。

- application：对应整个 Web 应用，该 Bean 只生成一个实例，这意味着，在整个 Web 应用内，程序每次请求该 Bean，得到的总是同一个实例。只有在 Web 应用中使用 Spring 时，该作用域才真正生效。
- websocket：在整个 WebSocket 的通信过程中，该 Bean 只生成一个实例。只有在 Web 应用中使用 Spring 时，该作用域才真正生效。

从上面 6 种作用域的描述中，我们可以得到常用的是 singleton(单例)和 prototype(原型)两种作用域，而其余 4 种均只有在 Web 应用中才有效。

### 6.4.4 单例和原型

单例(singleton)和原型(prototype)是 Bean 的两种常用作用域。如果不指定 Bean 的作用域，Spring 默认使用 singleton 作用域。

singleton 作用域的 Bean 实例是同一个共享实例，可以重复使用。而 prototype 作用域的 Bean 实例都是全新产生的实例，相当于 new 操作。而实例的创建和销毁等工作都会增加系统的开销，因此，应该尽量避免将 Bean 设置成 prototype 作用域。

下面通过代码来演示上面两种作用域下的 Bean。

修改配置文件 beans.xml，代码如下：

```
<?xml version="1.0" encoding="UTF-8"?>
<beans xmlns="http://www.springframework.org/schema/beans"
        xmlns:xsi="http://www.w3.org/2001/XMLSchema-instance"
        xsi:schemaLocation="http://www.springframework.org/schema/beans
                            http://www.springframework.org/schema/beans/spring-beans.xsd">
    <!--配置一个单例 singleton 实例-->
    <bean id="userServiceImpl" class="com.bailiban.service.impl.UserServiceImpl">
        <property name="userDao" ref="userDaoImpl"/>
    </bean>
    <!--配置一个原型 prototype 实例-->
    <bean id="userServiceImpl2" class="com.bailiban.service.impl.UserServiceImpl"
    scope="prototype">
        <property name="userDao" ref="userDaoImpl"/>
    </bean>
    <bean name="userDaoImpl" class="com.bailiban.dao.impl.UserDaoImpl"></bean>
</beans>
```

测试代码如下：

```
@Test
public void testScope() {
    ApplicationContext context = new ClassPathXmlApplicationContext("beans.xml");
    // 单例模式下的 Bean
```

```
        UserService userService_singleton1 = context.getBean("userServiceImpl", UserService.class);
        UserService userService_singleton2 = context.getBean("userServiceImpl", UserService.class);
        // 原型模式下的 Bean
        UserService userService_prototype1 = context.getBean("userServiceImpl2", UserService.class);
        UserService userService_prototype2 = context.getBean("userServiceImpl2", UserService.class);
        System.out.println(userService_singleton1 == userService_singleton2);
        System.out.println(userService_prototype1 == userService_prototype2);
    }
```

程序执行结果如下：

```
true
false
```

从上面的运行结果可以看出，对于 singleton 作用域下的 Bean，每次请求该 id 的 Bean，都将返回同一个共享实例，因而两次获取的 Bean 实例完全相同；但对 prototype 作用域的 Bean，每次请求该 id 的 Bean 都将产生新的实例，因此两次请求获得的 Bean 实例不相同。

## 6.4.5　Bean 的属性

在 Spring 中，Bean 的属性值有两种注入方式：即设置值的注入方式和构造方法注入方式，在 Spring 的配置文件中使用<property>元素驱动 Spring 执行 setter 方法和<constructor-arg>元素驱动 Spring 执行带参数的构造器。

本书中主要讲解使用设置值的方式，即采用<property>标记来进行属性的注入。

### 1. 简单值的注入

将简单的值注入到 Bean 中是非常容易的，首先定义一个类，通过暴露 setter 方法来声明它所需要设置的属性。

为了简化 getter、setter 方法，在 pom 文件中加入 lombok 插件依赖。依赖代码如下：

```
<dependency>
    <groupId>org.projectlombok</groupId>
    <artifactId>lombok</artifactId>
    <version>1.18.12</version>
</dependency>
```

Bean 代码如下所示：

```
package com.bailiban.pojo;


import lombok.Getter;
import lombok.Setter;
```

```
/**
 * @author zhulang
 * @Classname SimpleBean
 * @Description 简单的 bean
 * @Date 2020/7/17 15:45
 */
@Getter
@Setter
public class SimpleBean {
    private String name;
    private int age;
    private float height;
    private boolean male;

    @Override
    public String toString() {
        return "姓名:" + name + "\n"
                + "年龄:" + age + "\n"
                + "身高:" + height + "\n"
                + "是否男性:" + male;
    }
}
```

其相应的 Spring 配置文件示例代码如下：

```
<?xml version="1.0" encoding="UTF-8"?>
<beans xmlns="http://www.springframework.org/schema/beans"
        xmlns:xsi="http://www.w3.org/2001/XMLSchema-instance"
        xsi:schemaLocation="http://www.springframework.org/schema/beans
                        http://www.springframework.org/schema/beans/spring-beans.xsd">
    <bean id="simpleBean" class="com.bailiban.pojo.SimpleBean">
        <property name="name" value="bailiban"/>
        <property name="age" value="24"/>
        <property name="height" value="180.0"/>
        <property name="male" value="true"/>
    </bean>
</beans>
```

通过在 Bean 标记下添加一个或多个<property>子节点来指定要设置的属性，property 元素中通过 value 属性配置参数值。对于简单的普通属性值，Spring 使用 XML 解析器解析出这些数据，然后利用 java.beans.PropertyEditor 完成类型转换：从 java.lang.String 类型转换为所需的参数值类型。例如上面的 male 的 value 值为 true，会自动转换成布尔类型。

### 2. 组件的注入

除了注入简单的值外，Spring 还允许注入 Bean 对象，通过<ref>标记来完成组件间的依赖。例如在前面 6.3 节中的 beans.xml 中就已经使用到，代码如下：

```
<?xml version="1.0" encoding="UTF-8"?>
<beans xmlns="http://www.springframework.org/schema/beans"
        xmlns:xsi="http://www.w3.org/2001/XMLSchema-instance"
        xsi:schemaLocation="http://www.springframework.org/schema/beans
                            http://www.springframework.org/schema/beans/spring-beans.xsd">
    <bean id="userServiceImpl" class="com.bailiban.service.impl.UserServiceImpl">
        <property name="userDao" ref="userDaoImpl"/>
    </bean>
    <bean name="userDaoImpl" class="com.bailiban.dao.impl.UserDaoImpl"></bean>
</beans>
```

userServiceImpl 实例中依赖 UserDaoImpl 实例，通过 ref 属性，即可将容器中 UserDaoImpl 实例作为调用 setter 方法的参数。

### 3. 集合的注入

很多时候，Bean 组件的属性可能不仅仅是单一的值，而是一个对象集合。在 Spring 中为一个 Bean 注入一个集合属性也很简单，可以选择<list>、<map>、<set>和<props>来描述 List、Map、Set 和 Properties 的实例，然后和简单值的注入方式一样进行属性的注入。

接下来的示例显示了如何注入最常用到的 List、Map 和 Properties 类型的集合属性。首先是定义一个类，代码如下所示：

```
package com.bailiban.pojo;

import lombok.Getter;
import lombok.Setter;

import java.util.List;
import java.util.Map;
import java.util.Properties;

/**
 * @author zhulang
 * @Classname CountryBean
 * @Description TODO
 * @Date 2020/7/17 18:07
 */
@Setter
@Getter
public class CountryBean {
    private List countryList;
    private Map capitalMap;
    private Properties content;

    public void display() {
```

```
            for (int i = 0; i < countryList.size(); i++) {
                String country = (String) countryList.get(i);
                String capital = (String) capitalMap.get(country);
                System.out.println("国家:" + country + ",首都:" + capital);
            }
            System.out.println(content);
        }
}
```

相应的 Spring 配置文件为：

```
<?xml version="1.0" encoding="UTF-8"?>
<beans xmlns="http://www.springframework.org/schema/beans"
        xmlns:xsi="http://www.w3.org/2001/XMLSchema-instance"
        xsi:schemaLocation="http://www.springframework.org/schema/beans
                            http://www.springframework.org/schema/beans/spring-beans.xsd">
    <bean id="countryBean" class="com.bailiban.pojo.CountryBean">
        <property name="countryList">
            <list>
                <value>中国</value>
                <value>美国</value>
                <value>俄罗斯</value>
            </list>
        </property>
        <property name="capitalMap">
            <map>
                <entry key="中国" value="北京"/>
                <entry key="美国" value="华盛顿"/>
                <entry key="俄罗斯" value="莫斯科"/>
            </map>
        </property>
        <property name="content">
            <props>
                <prop key="象征">国旗</prop>
                <prop key="意义">国歌</prop>
            </props>
        </property>
    </bean>
</beans>
```

使用<list>注入一个 List 的实例，每一个条目可以通过<value>或是<ref>标记来指定。类似的，可以使用<map>注入一个 Map 的实例，每一个条目则是通过<entry>来指定，并且每个条目都由 key 或 key-ref 属性来指定键，由 value 属性或 value-ref 来指定值。使用<props>元素注入 Properties 类型的参数值，Properties 类型是一种特殊的类型，其 key 和 value 都只能是字符串，每个条目则是通过<prop>元素指定，key 属性指定 key 的值，<prop>元素的内容指定 value 的值。

### 4. 自动装配(autowire)

明确指定组件之间的依赖关系是 Spring 的 BeanFactory 推荐的用法，因为它允许按名称引用特定 bean 组件，即便有多个 bean 是相同的数据类型也不会混淆，并且通过 XML 或者别的 bean 工厂描述形式，各个 bean 之间的依赖关系一目了然，非常清晰。

除了明确的指定依赖关系外，Spring 的 BeanFactory 还提供了另外一种依赖决议的方法：自动装配(autowire)。如果一个 bean 被声明为自动装配，则 BeanFactory 会将它所管理的 bean 组件按照要求的依赖关系进行自动匹配，从而完成对象的装配。因为不需要明确指定某个依赖对象，所以可以带来很多方便，但前提条件是对象关系无歧义时才能完成自动装配。

Spring 最常用的自动装配模式有 byName(通过命名)、byType(通过类型)两种。

(1) byName：在<bean>元素中使用 autowire 属性，值设置为 byName。Spring 容器查找容器中的全部 Bean，找出其 id 与 setter 方法名去掉 set 前缀，并小写首字母后同名的 Bean 来完成注入。如果没有找到匹配的 Bean 实例，则 Spring 不会进行任何注入。

(2) byType：在<bean>元素中使用 autowire 属性，值设置为 byType。Spring 容器查找容器中的全部 Bean，如果正好有一个 Bean 类型与 setter 方法的形参类型匹配，就自动注入这个 Bean；如果找到多个这样的 Bean，就抛出一个异常；如果没有找到这样的 Bean，则什么都不会发生，setter 方法不会被调用。

自动装配可以减少配置文件的工作量，但降低了依赖关系的透明性和清晰性。本书并不推荐自动装配，所以就不在此进行额外演示。

# 6.5 Bean 装配中注解的使用

通过上面的学习，对于如何使用 XML 的方式装配 Bean，相信大家已经有了清晰的认识。但是很多时候我们会考虑使用注解的方式装配 Bean。使用注解的方式可以减少 XML 的配置，有利于开发人员更加高效地完成系统的开发。

## 6.5.1 使用@Component 注解装配 Bean

首先定义一个 POJO，我们将之前的 SimpleBean 代码改造一下，如下所示：

```
package com.bailiban.pojo;

import lombok.Setter;
import org.springframework.beans.factory.annotation.Value;
import org.springframework.stereotype.Component;
```

```
/**
 * @author zhulang
 * @Classname SimpleBean
 * @Description  简单的 bean
 * @Date 2020/7/17 15:45
 */
@Setter
@Component("simpleBean")
public class SimpleBean {
    @Value("厚溥")
    private String name;
    @Value("20")
    private int age;
    @Value("180.0")
    private float height;
    @Value("true")
    private boolean male;

    @Override
    public String toString() {
        return "姓名:" + name + "\n"
                + "年龄:" + age + "\n"
                + "身高:" + height + "\n"
                + "是否男性:" + male;
    }
}
```

代码中加粗的注解解释如下：

- @Component：代表 Spring IoC 会把这个类扫描生成 Bean 实例，而其中的 value 属性代表这个类在 Spring 中的 id，相当于 XML 方式定义的 Bean 的 id，也可以简写成@Component("simpleBean")，或者直接写成@Component，那么其 id 就是类名首字母小写。
- @Value：代表值的注入，这里只是注入一些简单值，注入的时候 Spring 会为其转化类型。

虽然定义好了要注入的类，但是 Spring 容器并不知道去哪里扫描，因此还需要创建一个配置类。代码如下：

```
package com.bailiban;

import org.springframework.context.annotation.ComponentScan;
import org.springframework.context.annotation.Configuration;

/**
```

```
 * @author zhulang
 * @Classname AppConfig
 * @Description Java 配置类
 * @Date 2020/7/20 16:04
 */
@Configuration
@ComponentScan
public class AppConfig {
}
```

这个类十分简单，加粗部分两个注解的含义如下。

- @Configuration：用于修饰一个配置类，等同于之前的 XML 配置文件。
- @ComponentScan：代表要扫描的包路径，默认是扫描当前包及其子包。

这样的话，前面定义的 com.bailiban.pojo 包下的 SimpleBean，属于配置类 com.bailiban 的子包路径下，会被 Spring IoC 容器扫描并装配进容器中。

使用注解方式装配 Bean，需要通过 AnnotationConfigApplicationContext 类来生成 Spring 容器。代码如下：

```
@Test
public void testSimpleBean() {
    // 使用 XML 方式创建 Spring 容器
    // ApplicationContext context = new ClassPathXmlApplicationContext("beans.xml");
    // 使用注解方式创建 Spring 容器
    ApplicationContext context = new AnnotationConfigApplicationContext(AppConfig.class);
    SimpleBean simpleBean = context.getBean("simpleBean", SimpleBean.class);
    System.out.println(simpleBean);
}
```

通过注解的方式的确大大简化了开发，但是依旧存在不足，其一，@ComponentScan 只是扫描所在包的 Java 类，但是更多的时候真正需要的是可以扫描我们指定的类；其二，上面只是注入了一些简单的值，而没有注入对象，但在现实开发中注入对象是常见的场景。下面先解决第一个问题，第二个问题会在后面小节中讲解。

@ComponentScan 存在着两个配置项：第一个是 basePackages，值可以是一个 Java 包的数组，Spring 会根据配置扫描对应的包和子包，将配置好的 Bean 装配进来。第二个是 basePackageClasses，值可以是多个类的数组，Spring 会根据配置的类所在的包，为包和子包扫描装配对应配置的 Bean。basePackages 的可读性会更好一些，因此建议优先选择使用它。例如指定扫描 pojo、service 包及其子包，代码如下所示：

```
@Configuration
@ComponentScan(basePackages = {"com.bailiban.pojo","com.bailiban.service"})
public class AppConfig {
}
```

### 6.5.2 使用@Autowired 注解完成自动装配

所谓自动装配，也就是让 Spring 自己发现对应的 Bean，自动完成装配工作的方式。此时会使用一个十分常用的注解@Autowired，这个时候 Spring 会根据类型去寻找定义的 Bean，然后将其注入。

下面，对之前的 UserServiceImpl 代码进行修改，如下所示：

```
@Service
public class UserServiceImpl implements UserService {
    @Autowired
    private UserDao userDao;

    @Override
    public void login() {
        userDao.login();
    }
}
```

这里我们使用了一个新的注解@Service，其本质和@Component 没有什么区别，都是让 Spring 容器来管理装配所在的类，只是为了更好地对应用进行分层处理。@Service 注解用在处理业务逻辑的类上，后面会用到的@Repository 用在持久层中，而@Component 注解则通用一些，没有特定标记。

上面代码中，我们不再使用 setter 注入或构造注入 UserDao，而是通过@Autowired 注解注入。这样就完成了自动注入的装配。

使用了@Autowired 注解，Spring 容器会认为一定要找到对应的 Bean 来注入这个字段，否则会抛出异常。可以通过@Autowired 的配置项 required 来改变它，设置为@Autowired(required = false)，这样 Spring 容器找不到对应的 Bean 类型，允许不注入。required 的默认值为 true。

### 6.5.3 @Qualifier 注解的使用

在上面我们讲到@Autowired 注解是按照要注入 Bean 的类型来注入对象。而在 Java 中接口可以有多个实现类，同样的抽象类也可以有多个实例化的类，这样就会造成通过类型获取 Bean 的不唯一，从而导致 Spring 容器无法获得唯一的实例化类。

为了解决这种歧义性，Spring 提供了@Qualifier 注解，它可以按照名称的方式进行查找。为了演示这种情况，我们再创建一个 UserDao 的实现类 UserDaoImpl2，代码如下所示。

```
@Repository
public class UserDaoImpl2 implements UserDao {
```

```
    @Override
    public void login() {
        System.out.println("登录成功！--UserDaoImpl2");
    }
}
```

修改 UserServiceImpl 代码，如下所示。

```
@Service
public class UserServiceImpl implements UserService {
    @Autowired
    @Qualifier("userDaoImpl2")
    private UserDao userDao;

    @Override
    public void login() {
        userDao.login();
    }
}
```

上面代码中，UserDao 有多个实现类，我们指定使用名称为 userDaoImpl2 这个实现类。这个时候 Spring 容器就不会再按照类型的方式注入，而是按照名称的方式注入，这样既能注入成功，而且也不存在歧义性。

## 6.5.4　@Bean 注解的使用

在前面我们通过@Component 或者特定的@Service 和@Repository 装配 Bean，但是这些注解只能作用在类上，不能注解到方法上。对于 Java 而言，大部分的开发都需要引入第三方的包(jar 文件)，而且往往并没有这些包的源码，这时候将无法为这些包的类加入@Component 注解，让它们变为开发环境的 Bean。

这个时候 Spring 提供了一个注解@Bean，它可以注解到方法上，将该方法的返回值定义成容器中的一个 Bean。

(1) 先创建一个 Bean，名为 MyBean。

```
public class MyBean {
    public MyBean() {
        System.out.println("MyBean Initializing");
    }
}
```

在 MyBean 类上，我们没有加上@Component 注解，可以将其当成第三方的包。

(2) 再在 AppConfig 配置类上注入 Bean 的代码，完整代码如下。

```
@Configuration
```

```
@ComponentScan(basePackages = {"com.bailiban.pojo", "com.bailiban.service", "com.bailiban.dao"})
public class AppConfig {
    @Bean
    public MyBean myBean() {
        return new MyBean();
    }
}
```

这样使用@Bean注解就可以将所修饰方法的返回值注入Spring容器中，在未指定Bean的名称时，默认为方法名首字母小写。也可以自行指定其名称，例如@Bean("mBean")，指定注入Spring容器的Bean名称为mBean。

# 6.6 Bean的生命周期

前面的章节介绍了Spring中Bean的相关属性等基础知识，一个Bean从定义到销毁都有一个生命周期。在Spring中，Bean的完整的生命周期包括定义、初始化、使用和销毁四个阶段，其中Bean的定义和使用在前面都有相关的内容，在这里就不再描述，接下来重点来描述Bean的初始化和销毁。

## 6.6.1 Bean的初始化

在Spring当中，Bean的初始化有两种方式：

(1) 使用init-method属性。

(2) 实现InitializingBean接口。

需要说明的是，使用第二种方式具有一定的侵入性，因为实现了容器特定的InitializingBean接口，bean对象就不是POJO了，不具备通用性，所以要尽量使用第一种方式，在这里不介绍第二种方法了，有兴趣的读者可自行阅读Spring的参考文档。

在第一种方式中，通过配置Bean的init-method属性来完成，具体步骤如下：

首先定义Bean对应的类MsgBean，添加一个无参的方法init，用它对Bean进行初始化。

```
@Setter
public class MsgBean {
    private String message;
    private Date date;

    /**
     * 初始化方法
     */
```

```
    public void init() {
        date = new Date();
        System.out.println("正在执行初始化方法 init……");
    }

    public void display() {
        System.out.println(date + ":" + message);
    }
}
```

然后在 Spring 的配置文件中设置 bean 的 init-method 属性：

```
<?xml version="1.0" encoding="UTF-8"?>
<beans xmlns="http://www.springframework.org/schema/beans"
       xmlns:xsi="http://www.w3.org/2001/XMLSchema-instance"
       xsi:schemaLocation="http://www.springframework.org/schema/beans
                           http://www.springframework.org/schema/beans/spring-beans.xsd">
    <bean id="msgBean" class="com.bailiban.pojo.MsgBean" init-method="init">
        <property name="message" value="Hello,Spring!"/>
    </bean>
</beans>
```

通过 BeanFactory 获取 msgbean 对象后，Spring 会自动在 msgbean 对象上调用 init-method 中指定的 init 方法进行初始化。

## 6.6.2　Bean 的销毁

在 Spring 当中，Bean 的销毁也有两种方式：

(1) 使用 destroy-method 属性。

(2) 实现 DisposableBean 接口。

使用第二种方式也具有侵入性，因为实现了容器特定的 DisposableBean 接口，和 Bean 的初始化一样，所以要尽量使用第一种方式，同样的对第二种方式也不作介绍了。

在第一种方式中，通过配置 Bean 的 destory-method 属性来完成，具体步骤如下：

首先定义 Bean 对应的类 msgBean，添加一个无参的方法 destory，用它对 Bean 进行销毁处理。

```
@Setter
public class MsgBean {
    private String message;
    private Date date;

    /**
     * 初始化方法
     */
```

```
    public void init() {
        date = new Date();
        System.out.println("正在执行初始化方法 init……");
    }

    /**
     * 销毁方法
     */
    public void destory() {
        message = "";
        date = null;
        System.out.println("正在执行销毁之前的方法 destroy……");
    }

    public void display() {
        System.out.println(date + ":" + message);
    }
}
```

然后在 Spring 的配置文件中设置 bean 的 destory-method 属性：

```
<?xml version="1.0" encoding="UTF-8"?>
<beans xmlns="http://www.springframework.org/schema/beans"
       xmlns:xsi="http://www.w3.org/2001/XMLSchema-instance"
       xsi:schemaLocation="http://www.springframework.org/schema/beans
                           http://www.springframework.org/schema/beans/spring-beans.xsd">
    <bean id="msgBean" class="com.bailiban.pojo.MsgBean" init-method="init"
     destroy-method="destory">
        <property name="message" value="Hello,Spring!"/>
    </bean>
</beans>
```

当通过 BeanFactory 对 msgbean 进行销毁时，Spring 会首先在 msgbean 对象上调用 destory-method 中的指定的 destory 方法进行处理。

### 6.6.3 使用注解定义 Bean 的初始化和销毁

在讲完使用 XML 方式定义 Bean 的初始化和销毁后，下面讲解如何使用注解的方式完成。代码如下。

```
@Configuration
@ComponentScan(basePackages = {"com.bailiban.pojo", "com.bailiban.service", "com.bailiban.dao"})
public class AppConfig {
    @Bean("myBean")
    public MyBean myBean() {
        return new MyBean();
```

```
    }

    @Bean(value = "msgBean", initMethod = "init", destroyMethod = "destory")
    public MsgBean msgBean() {
        return new MsgBean();
    }
}
```

@Bean 的配置项中包含 4 个配置项。

- name 或 value：一个字符串数组，允许配置多个 BeanName。
- autowire：标志是否是一个引用的 Bean 对象，默认值是 Autowire.NO。
- initMethod：自定义初始化方法。
- destroyMethod：自定义销毁方法。

这样通过@Bean 注解的 initMethod 和 destroyMethod 就完成了 Bean 的初始化和销毁方法的定义。

## 单元小结

- 控制反转有两种主要的方式：依赖查找和依赖注入，其中依赖注入又主要分为设值注入方式和构造方法注入方式。
- ApplicationContext 是 Spring 容器最常用的接口，因此 Spring 容器也称为 Spring 上下文。
- Spring 默认使用 singleton 作用域，此作用域的 Bean 实例是同一个共享实例，可以重复使用。
- 可以使用@Autowired 和@Qualifier 注解一起完成多个同类型 Bean 情况下的自动注入。
- Spring 中的 bean 的完整生命周期包括定义、初始化、使用、销毁。

1. 下列有关依赖注入说法错误的是(　　)。
   A. 依赖注入一般包括两种方式，即设值注入和构造方法注入
   B. Spring 只支持设值注入方式一种
   C. 设值注入方式使用得多一些
   D. 依赖的注入一般情况下由容器完成，不是应用代码完成的

2. 下面关于 IoC 的理解，正确的两项是(　　)。

A. 控制反转

B. 对象被动接受依赖类

C. 对象主动接受依赖类

D. 一定要有接口

3. 下面关于在 Spring 中配置 Bean 的 id 属性的说法，正确的两项是(　　)。

A. id 属性是必须，没有 id 属性就会报错

B. id 属性不是必须的，可以没有

C. id 属性的值可以重复

D. id 属性的值不可以重复

4. IoC 自动装载方法的两项是(　　)。

A. byName　　B. byType

C. constructor　　D. byMethod

5. 下面关于在 Spring 中配置 Bean 的 init-method 的说法，正确的是(　　)。

A. init-method 是在最前面执行的

B. init-method 在构造方法后，依赖注入前执行

C. init-method 在依赖注入之后执行

D. init-method 在依赖注入之后，构造函数之前执行

上机目标

- 使用 Idea 编写 Spring 应用程序。
- 掌握 Spring 配置文件的编写。

上机练习

◆第一阶段◆

练习 1：beans.xml 配置文件的编写

【问题描述】

使用 Spring 注入简单的属性以及 bean 的引用，并使用该 bean 进行编程。

【问题分析】

一个 Spring 应用程序的源代码主要包括两部分：JavaBean 和 Bean 的配置文件，配置文件的头信息可以使用 Idea 进行创建。

【参考步骤】

请参照本章 spring-demo 工程的步骤代码。

◆第二阶段◆

### 练习 2：使用 XML 和 Java 类配置混合编写 Spring 应用程序

【问题分析】

(1) 除了使用 XML 配置外，很多开发者更偏向于使用注解方式来配置依赖关系。

(2) 不论使用 XML 还是注解，目的都是简化代码，同时提高代码的健壮性，因此在实际项目中可能会混合使用。

【参考步骤】

(1) 首先创建一个 Maven 工程，代码参照本单元的 spring-demo 工程。

(2) 如果以 XML 配置为主，Java 配置为辅，就需要让 XML 配置能加载 Java 类配置。参考代码如下：

```
<?xml version="1.0" encoding="UTF-8"?>
<beans xmlns="http://www.springframework.org/schema/beans"
        xmlns:xsi="http://www.w3.org/2001/XMLSchema-instance"
        xsi:schemaLocation="http://www.springframework.org/schema/beans
                        http://www.springframework.org/schema/beans/spring-beans.xsd">
    <bean id="msgBean" class="com.bailiban.pojo.MsgBean" init-method="init" destroy-method="destory">
        <property name="message" value="Hello,Spring!"/>
    </bean>
    <!--加载 Java 配置类-->
    <bean class="com.bailiban.AppConfig"></bean>
</beans>
```

此时创建 Spring 容器时，通过创建 ClassPathXmlApplicationContext 对象作为 Spring 容器。Spring 会先加载这份 XML 配置文件，再根据这份 XML 配置文件的指示去加载指定的 Java 配置类。

(3) 如果以 Java 类配置为主，就需要让 Java 配置类能加载 XML 配置。这就需要借助于@ImportResource 注解，这个注解可修饰 Java 配置类，用于导入指定的 XML 配置文件。参考代码如下：

```
@Configuration
@ImportResource("classpath:/beans.xml")
```

```
public class AppConfig {
......
}
```

此时创建 Spring 容器时，通过创建 AnnotationConfigApplicationContext 对象作为 Spring 容器。Spring 会先加载这个配置类，再根据这个配置类的指示加载指定的 XML 配置文件。

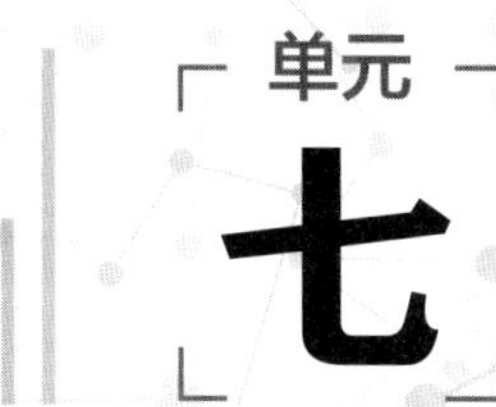


# Spring AOP编程

## 课程目标

- ❖ 了解 AOP 的基本概念
- ❖ 了解 Spring 中的 AOP
- ❖ 了解 AspectJ 的使用
- ❖ 了解 AOP 代理

**简介**

AOP(Aspect-Oriented Programming，面向切面编程)，作为面向对象编程的一种补充，已经成为一种比较成熟的编程方式。使用 AOP，可以将处理某些与程序功能无关但又是必须的代码注入主程序，如异常处理、日志、安全、事务等，通常主程序的主要目的并不在于处理这些切面(aspect)。

Spring AOP 是一种非侵略性的、轻型的 AOP 框架，无需使用预编译器或其他的元标签，便可以在 Java 程序中使用它，大大简化了 AOP 的开发复杂度，也降低了 AOP 的开发难度。

## 7.1 AOP 的基本概念

软件开发领域的一个核心问题就是如何能够更好地满足关注点分离 (separation of concerns)原则。这个原则表达了代码和开发过程的一个最为重要的特性，但遗憾的是它只是一个原则，其中并没有告诉我们如何做才能满足这一原则。人们在寻求能够满足这一原则的实现技术上进行了很多探索，也取得了许多成就。其中，AOP 就是一项新的技术。

AOP 的概念是由 Xerox PARC 研究中心的研究人员首先提出的，其目的是通过提供一些方法和技术，把问题领域分解成一系列的功能性组件(functional component)和一系列横跨多个功能性组件的切面(aspect)，然后组合这些组件和切面，获得系统的实现。

AOP，即面向切面编程，它可以使程序开发人员更好地将不应该彼此粘合在一起的功能代码分离开来。

开发人员在编写应用程序时，通常会编写两种代码：一种是与业务系统相关的代码，另一种是与业务系统关系不大但又是必须的辅助性代码，比如异常处理、日志、安全、事务处理等。通常在编写代码时，这两种代码是写在一起的，这样在程序中到处充斥着相同或类似的代码，不利于对程序的维护。而使用 AOP，就不用在业务逻辑代码中编写与业务系统功能关系不大的代码，将业务逻辑代码与辅助性代码进行分离，降低了这两种代码的耦合度，从而达到了易于维护和重用的目的。

## 7.2 AOP 的实现原理

### 7.2.1 代理模式简介

AOP 的实现是基于代理模式，我们先来了解一下什么是代理模式。

代理模式是指为其他对象提供一种代理，以控制对这个对象的访问。在某些情况下，一个客户不想或者不能直接引用另一个对象，这时就需要代理对象在客户端和目标对象之间起到中介的作用。

代理模式一般涉及的角色有：

- **抽象角色：**声明真实对象和代理对象的共同接口或者抽象类。
- **代理角色：**代理对象角色内部含有对真实对象的引用，从而可以操作真实对象，同时代理对象提供与真实对象相同的接口，以便在任何时刻都能代替真实对象。同时，代理对象可以在执行真实对象操作时，附加其他的操作，相当于对真实对象进行封装。
- **真实角色：**代理角色所代表的真实对象，是最终要引用的对象。

### 7.2.2 代理模式示例

前面已经介绍了一些概念，下面举一个例子说明代理模式是如何工作的。

比如，现在某人要卖房子，他可以自己卖，也可以找一家中介公司来帮着卖；如果是找中介公司帮着卖房子，中介公司会收取一定的费用，根据上面的描述我们来创建对应的类结构：

- **Seller(销售者)接口**，包含 sellHouse 方法。
- **Me(我)类**，实现了 Seller 接口，表示我可以卖房子。
- **AgentCompany(中介公司)类**，实现了 Seller 接口，表示中介公司也可以卖房子。

先在 Idea 中创建一个基本的 Maven 项目，名为 springaop-demo，可以勾选模板 maven-archetype-quickstart。创建好 Maven 项目后，在 pom 文件中添加 Spring 相关依赖坐标，代码如下所示。

1. pom.xml

```
<?xml version="1.0" encoding="UTF-8"?>
<project xmlns="http://maven.apache.org/POM/4.0.0"
  xmlns:xsi="http://www.w3.org/2001/XMLSchema-instance"
  xsi:schemaLocation="http://maven.apache.org/POM/4.0.0
```

```
        http://maven.apache.org/xsd/maven-4.0.0.xsd">
        <modelVersion>4.0.0</modelVersion>
        <groupId>com.bailiban</groupId>
        <artifactId>springaop-demo</artifactId>
        <version>1.0-SNAPSHOT</version>

        <name>springaop-demo</name>

        <properties>
            <project.build.sourceEncoding>UTF-8</project.build.sourceEncoding>
            <maven.compiler.source>1.8</maven.compiler.source>
            <maven.compiler.target>1.8</maven.compiler.target>
        </properties>

        <dependencies>
            <dependency>
                <groupId>junit</groupId>
                <artifactId>junit</artifactId>
                <version>4.11</version>
                <scope>test</scope>
            </dependency>
            <dependency>
                <groupId>org.springframework</groupId>
                <artifactId>spring-context</artifactId>
                <version>5.2.7.RELEASE</version>
            </dependency>
            <dependency>
                <groupId>org.projectlombok</groupId>
                <artifactId>lombok</artifactId>
                <version>1.18.12</version>
            </dependency>
        </dependencies>
    </project>
```

上面 pom 文件中依赖坐标和上一章节 spring-demo 工程中的 pom 文件依赖坐标其实一模一样。因为 spring-context 依赖本身是依赖于 spring-beans、spring-core、spring-aop 和 spring-expression 这四个 jar 包的，所以就无须再重复引入 spring-aop 依赖坐标。可以从 Idea 的 Maven 依赖结构图进行查看，如图 7-1 所示。

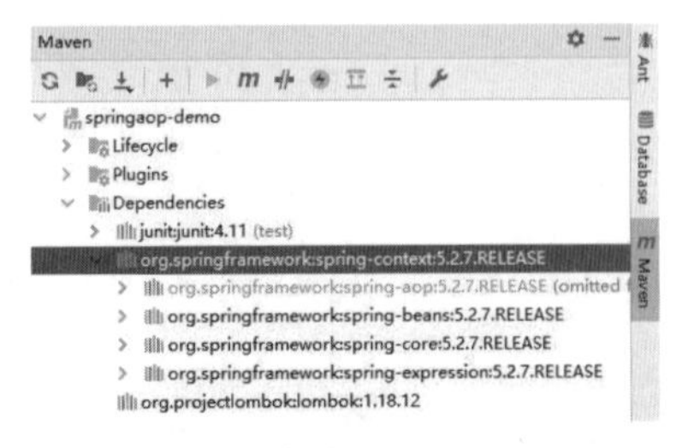


图 7-1

### 2. Seller 接口

```
package com.bailiban.aop;

/**
 * @author zhulang
 * @Classname Seller
 * @Description 销售者
 * @Date 2020/7/23 11:24
 */
  public interface Seller {
    /**
     * @param money 房子的售价
     */
    void getMoney(double money);
}
```

### 3. Me 实现类

表示房主自己作为销售方进行卖房。

```
package com.bailiban.aop;

/**
 * @author zhulang
 * @Classname Me
 * @Description 房主
 * @Date 2020/7/23 11:27
 */
public class Me implements Seller {
    @Override
    public void getMoney(double money) {
        System.out.println("收到了" + money);
    }
}
```

### 4. AgentCompany 实现类

表示中介公司作为销售方进行卖房。

```
package com.bailiban.aop;

/**
 * @author zhulang
 * @Classname AgentCompany
 * @Description 中介公司
 * @Date 2020/7/23 11:29
 */
public class AgentCompany implements Seller {
```

```
        @Override
        public void getMoney(double money) {
            // 中介公司收取了 5%的中介费
            System.out.println("收到了" + money);
            System.out.println("其中包含中介费" + money * 0.05);
        }
}
```

如果要实现“我”通过房产中介公司卖房，则需要对上面代码进行修改，如下所示。

```
package com.bailiban.aop;

/**
 * @author zhulang
 * @Classname AgentCompany
 * @Description  中介公司
 * @Date 2020/7/23 11:29
 */
public class AgentCompany implements Seller {
    private Seller seller = new Me();

    @Override
    public void getMoney(double price) {
        System.out.println("我委托中介公司帮我卖房");

        if (price > 0) {
            System.out.println("中介公司给了卖房款");
            //中介公司收取 5%手续费
            price = price * (1 - 0.05);
            //我收到了卖房款
            seller.getMoney(price);
        } else {
            System.out.println("房子没卖出去");
            seller.getMoney(0);
        }
    }
}
```

从修改过的 AgentCompany 类可以看出，AgentCompany 类中有一个 Seller 属性，是通过 Me 类来创建的，AgentCompany 就是代理模式中的代理角色，而 Me 是代理模式中的真实角色。

上面是我们生活中的例子，下面看看如何在现在的程序中使用代理模式。现在有一个用户的业务接口，一个用户的业务实现类，如果要在不修改代码的情况下给其中的批量删除方法加上事务。

### 5. UserService 用户的业务接口

```
package com.bailiban.service;

import java.util.List;

/**
 * @author zhulang
 * @Classname UserService
 * @Description  用户的业务接口
 * @Date 2020/7/24 9:55
 */
public interface UserService {
    /**
     * 批量删除用户
     *
     * @param idList  用户编号的列表
     */
    public void batchDelete(List<Integer> idList);
}
```

### 6. UserServiceImpl 用户业务实现类

```
package com.bailiban.service.impl;

import com.bailiban.service.UserService;

import java.util.List;

/**
 * @author zhulang
 * @Classname UserServiceImpl
 * @Description  用户业务的实现类
 * @Date 2020/7/24 9:56
 */
public class UserServiceImpl implements UserService {
    @Override
    public void batchDelete(List<Integer> idList) {
        // 具体代码略
    }
}
```

### 7. UserServiceTransactionProxy 用户业务类的事务代理类

因为代理模式允许在代理角色执行之前和之后做操作，这个要求可以通过代理模式来实现。可以创建一个类 UserServiceProxy 来完成代理的任务。

```
package com.bailiban.service.impl;

import com.bailiban.service.UserService;

import java.util.List;

/**
 * @author zhulang
 * @Classname UserServiceTransactionProxy
 * @Description 用户业务类的事务代理类
 * @Date 2020/7/24 9:57
 */
public class UserServiceTransactionProxy implements UserService {
    private UserService userService = new UserServiceImpl();

    @Override
    public void batchDelete(List<Integer> idList) {
        System.out.println("事务处理:开启事务");
        try {
            userService.batchDelete(idList);
            System.out.println("事务处理:提交事务");
        } catch (Exception e) {
            System.out.println("事务处理:回滚事务");
            throw e;
        }
    }
}
```

### 8. UserServiceLoggingProxy 用户业务类的日志代理类

目的已经达到了，在不修改 UserServiceImpl 代码的情况下，我们加上了事务；同样的，通过代理模式，可以在不修改代码的情况下给类加上日志、权限控制等功能。

```
package com.bailiban.service.impl;

import com.bailiban.service.UserService;

import java.util.List;

/**
 * @author zhulang
 * @Classname UserServiceLoggingProxy
 * @Description TODO
 * @Date 2020/7/24 10:02
 */
public class UserServiceLoggingProxy implements UserService {
    private UserService userService = new UserServiceTransactionProxy();
```

```
    @Override
    public void batchDelete(List<Integer> idList) {
        System.out.println("日志记录:进入批量删除方法");
        try {
            userService.batchDelete(idList);
            System.out.println("日志记录:批量删除成功");
        } catch (Exception e) {
            System.out.println("日志记录:批量删除失败");
            throw e;
        }
    }
}
```

### 9. 编写测试整个代理机构的类

UserServiceLoggingProxy 代理了 UserServiceTransactionProxy，在原有事务的基础上加了日志功能。下面编写一个类对整个代理结构进行测试。

```
package com.bailiban;

import com.bailiban.service.UserService;
import com.bailiban.service.impl.UserServiceLoggingProxy;
import org.junit.Test;

/**
 * Unit test for simple App.
 */
 public class AppTest {
    @Test
    public void testProxy() {
        UserService userService = new UserServiceLoggingProxy();
//没有具体的业务逻辑，所以没有传递具体参数
        userService.batchDelete(null);
    }
}
```

控制台显示结果，如图 7-2 所示。

```
✔ Tests passed: 1 of 1 test – 9 ms
"D:\Program Files\Java\jdk1.8.0_221\bin\java.exe" ...
日志记录:进入批量删除方法
事务处理:开启事务
事务处理:提交事务
日志记录:批量删除成功

Process finished with exit code 0
```

图 7-2

通过上面的案例可以知道，代理模式就是在不修改代码的情况下，在方法调用的过程

中加上新的功能。Spring 支持两种代理的方式，一种是 JDK 动态代理，另一种是 CGLIB 动态代理，接下来介绍这两种代理方式是如何工作的。

## 7.2.3 JDK 动态代理

前面一节已经使用编写代码的方式实现了代理模式，这种方式对于一个项目显得比较复杂，要多写很多类才能达到增加功能的目的。对于代理模式，JDK 提供了更加简便的方式来实现代理模式，叫做 JDK 动态代理。使用 JDK 动态代理必须要使用接口，动态代理主要靠 java.lang.reflect.Proxy 类和 java.lang.reflect. InvocationHandler 接口来实现，Proxy 类主要负责创建代理类，InvocationHandler 接口主要负责在方法调用前后加上什么功能。JDK 文档给出了创建接口 Foo 的代理实例的方式。

```
Foo f = (Foo) Proxy.newProxyInstance(Foo.class.getClassLoader(),
                                      new Class[] { Foo.class },
                                      handler);
```

下面将上一节的案例改成使用 JDK 动态代理的方式，看看区别在哪里；其中 UserService 和 UserServiceImpl 两个类可以不做任何修改。

### 1. TransactionInvocationHandler 事务的调用处理器

```
package com.bailiban.aop;

import java.lang.reflect.InvocationHandler;
import java.lang.reflect.Method;
import java.util.Arrays;

/**
 * @author zhulang
 * @Classname TransactionInvocationHandler
 * @Description 事务的调用处理器
 * @Date 2020/7/28 10:48
 */
public class TransactionInvocationHandler implements InvocationHandler {
    /**
     * 被代理对象
     */
    private Object target;

    public TransactionInvocationHandler(Object target) {
        super();
        this.target = target;
    }
```

```
    @Override
    public Object invoke(Object proxy, Method method, Object[] args)
                            throws Throwable {
        Object returnValue = null;
        System.out.println("事务处理:开启事务");
        System.out.println("方法声明于接口:" + method.getDeclaringClass());
        System.out.println("被代理对象属于类:" + target.getClass());
        System.out.println("方法的参数为:" + Arrays.toString(args));
        try {
            returnValue = method.invoke(target, args);
            System.out.println("事务处理:提交事务");
        } catch (Exception e) {
            System.out.println("事务处理:回滚事务");
            throw e;
        }
        return returnValue;
    }
}
```

TransactionInvocationHandler 的关键语句是 returnValue = method.invoke(target, args)，这是通过反射的方式来调用被代理对象的方法，args 是方法调用的参数，method 是被代理的方法，returnValue 是方法调用之后的返回值。如果通过 TransactionInvocationHandler 来创建代理对象，代理对象在调用任何方法的时候，都会执行 invoke 方法里面的内容，遇到 method.invoke 方法才会去调用被代理对象。通过这种方式，给被代理对象的所有方法都加上 invoke 方法中写的代码，根据这种思路，LoggingInvocationHandler 的代码也基本类似。

### 2. LoggingInvocationHandler 日志的调用处理器

```
package com.bailiban.aop;

import java.lang.reflect.InvocationHandler;
import java.lang.reflect.Method;
import java.util.Arrays;

/**
 * @author zhulang
 * @Classname LoggingInvocationHandler
 * @Description 日志的调用处理器
 * @Date 2020/7/28 11:10
 */
public class LoggingInvocationHandler implements InvocationHandler {
    /**
     * 被代理对象
     */
    private Object target;
```

```
    public LoggingInvocationHandler(Object target) {
        super();
        this.target = target;
    }

    @Override
    public Object invoke(Object proxy, Method method, Object[] args)
                    throws Throwable {
        Object returnValue = null;
        System.out.println("日志记录:进入" + method.getName() + "方法");
        System.out.println("方法声明于接口:" + method.getDeclaringClass());
        System.out.println("被代理对象属于类:" + target.getClass());
        System.out.println("方法的参数为:" + Arrays.toString(args));
        try {
            returnValue = method.invoke(target, args);
            System.out.println("日志记录:调用" + method.getName() + "方法成功");
        } catch (Exception e) {
            System.out.println("日志记录:调用" + method.getName() + "方法失败");
            throw e;
        }
        return returnValue;
    }
}
```

两个调用处理器都已经写好了，下面写一个类对动态代理进行测试。

### 3. 测试方法

```
@Test
public void reflectProxy() {
    // 创建事务的调用处理器
    TransactionInvocationHandler transactionInvocationHandler = new
            TransactionInvocationHandler(new UserServiceImpl());
    // 创建代理对象
    UserService userService = (UserService) Proxy
                            .newProxyInstance(UserService.class.getClassLoader(),
                            new Class[]{UserService.class},
                            transactionInvocationHandler);
    // 创建日志的调用处理器
    LoggingInvocationHandler loggingInvocationHandler = new
            LoggingInvocationHandler(userService);
    // 创建代理对象
    userService = (UserService) Proxy.newProxyInstance(UserService.class.getClassLoader(),
                new Class[]{UserService.class}, loggingInvocationHandler);
    // 创建批量删除方法的参数
    List<Integer> idList = new ArrayList<Integer>();
```

```
        idList.add(1);
        idList.add(2);
        // 调用方法
        userService.batchDelete(idList);
    }
```

测试后的输出结果如图 7-3 所示。

```
✔ Tests passed: 1 of 1 test – 17 ms
"D:\Program Files\Java\jdk1.8.0_221\bin\java.exe" ...
日志记录:进入batchDelete方法
方法声明于接口:interface com.bailiban.service.UserService
被代理对象属于类:class com.sun.proxy.$Proxy2
方法的参数为:[[1, 2]]
事务处理:开启事务
方法声明于接口:interface com.bailiban.service.UserService
被代理对象属于类:class com.bailiban.service.impl.UserServiceImpl
方法的参数为:[[1, 2]]
事务处理:提交事务
日志记录:调用batchDelete方法成功

Process finished with exit code 0
```

图 7-3

由上面的案例可以看出，JDK 动态代理可以代理多个类的多个方法，而且可以动态增加调用处理器，完成多个功能，并且不需要修改代码，如果对接口名、类名、方法名、参数类型进行判断，还可以有选择地代理部分方法，Spring 默认使用 JDK 动态代理的方式进行面向切面编程。

## 7.2.4　CGLIB 动态代理

CGLIB(Code Generation Library)是一个强大的高性能的代码生成包。它被许多 AOP 的框架广泛使用，例如，Spring AOP 和 dynaop 为它提供方法的 interception(拦截)。流行的 ORM 框架 Hibernate 也使用 CGLIB 来实现延迟加载。

代理为控制要访问的目标对象提供了一种途径。当访问对象时，它引入了一个间接的层。JDK 从 1.3 版本开始，就引入了动态代理，并且经常被用来动态地创建代理。JDK 的动态代理用起来非常简单，但它有一个限制，就是使用动态代理的对象必须实现一个或多个接口。如果想代理没有实现接口的继承的类，该怎么办？Spring 也可以使用 CGLIB 代理，在需要代理类而不是代理接口的时候，Spring 会自动切换为使用 CGLIB 代理。但 Spring 推荐使用面向接口编程，因此业务对象通常都会实现一个或多个接口，此时默认将使用 JDK 动态代理，但也可以强制使用 CGLIB 代理。只需要在 XML 配置文件中加入如下配置即可。

```
<aop:config proxy-target-class="true"/>
```

使用 CGLIB 主要靠 MethodInterceptor 接口和 Enhancer 类，MethodInterceptor 接口的作用和 JDK 动态代理中的 InvocationHandler 接口一样，是负责加入进来的代码，Enhancer

类和 Proxy 类作用也一样，用来产生代理对象。

下面将前面批量删除用户的例子改写为使用 CGLIB 代理。

### 1. UserService

UserService 现在可以是一个类，不需要实现任何接口。

```
package com.bailiban.cglib;

import java.util.List;

/**
 * @author zhulang
 * @Classname UserService
 * @Description 用户业务的实现类
 * @Date 2020/7/28 16:18
 */
public class UserService {
	public void batchDelete(List<Integer> idList) {
		System.out.println("正在进行批量删除");
		// 代码略
	}
}
```

### 2. TransactionMethodInterceptor

```
package com.bailiban.cglib;

import org.springframework.cglib.proxy.MethodInterceptor;
import org.springframework.cglib.proxy.MethodProxy;

import java.lang.reflect.Method;

/**
 * @author zhulang
 * @Classname TransactionMethodInterceptor
 * @Description 事务的方法拦截器
 * @Date 2020/7/28 16:26
 */
public class TransactionMethodInterceptor implements MethodInterceptor {
	@Override
	public Object intercept(Object proxy, Method method, Object[] args,
						MethodProxy methodProxy) throws Throwable {
		Object returnValue = null;

		try {
			System.out.println("事务处理:开启事务");
			returnValue = methodProxy.invokeSuper(proxy, args);
```

```
                System.out.println("事务处理:提交事务");
            } catch (Exception e) {
                System.out.println("事务处理:回滚事务");
                throw e;
            }
            return returnValue;
        }
    }
```

### 3. 测试方法

```
@Test
public void cglibProxy() {
    Enhancer enhancer = new Enhancer();
    // 设置为 UserService 做代理
    enhancer.setSuperclass(UserService.class);
    // 设置回调，TransactionMethodInterceptor 相当于 JDK 动态代理中的 InvocationHandler
    enhancer.setCallback(new TransactionMethodInterceptor());
    // 创建代理对象
    UserService userService = (UserService) enhancer.create();
    // 创建批量删除方法的参数
    List<Integer> idList = new ArrayList<Integer>();
    idList.add(1);
    idList.add(2);
    // 调用方法
    userService.batchDelete(idList);
}
```

测试的显示结果如图 7-4 所示。

```
✔ Tests passed: 1 of 1 test – 544 ms
"D:\Program Files\Java\jdk1.8.0_221\bin\java.exe" ...
事务处理:开启事务
正在进行批量删除
事务处理:提交事务

Process finished with exit code 0
```

图 7-4

## 7.2.5　AOP 的关键概念

学习 AOP，关键在于理解 AOP 的思想，以便正确使用 AOP。对于 AOP 中的诸多概念，只需要理解四个最重要的概念，它们是 Joinpoint、Pointcut、Advice、Aspect。

(1) Joinpoint(连接点)

连接点指的是程序运行中的某个阶段点，如方法的调用、异常抛出等。在 Spring AOP 中，连接点总是方法的调用。

(2) Pointcut(切入点)

可以插入增强处理的连接点。比如一个类中有多个方法连接点，但只有某个方法满足指定要求，这个连接点将被添加增强处理，该连接点也就变成了切入点。如何使用表达式来定义切入点是 AOP 的核心？Spring 默认使用 AspectJ 切入点语法。

(3) Advice(通知/增强处理)

Advice 是某个连接点所采用的处理逻辑，也就是向连接点注入的代码。处理有 around、before 和 after 等类型。

(4) Aspect(切面)

切面是通知和切点的集合，通知和切点共同定义了切面的全部功能——它是什么，在何时何处完成其功能。

## 7.3 Spring 中的 AOP

AOP 是一个概念，并没有设定具体语言的实现，它能克服那些只有单继承特性语言的缺点(如 Java)，目前 AOP 的具体实现有很多，其中 Spring 也有自己的实现，它本身不仅是一个 AOP 框架，还提供了一些现成的切面，比如"事务管理"切面，从某种意义上说这些切面本身比 AOP 框架更为重要。

Spring AOP 将 AOP 支持与 Spring 的轻量级容器基础设施整合在一起，将 Spring 提供服务与 AOP 组合起来。当然，也可以在 Spring 的 Bean 工厂或应用程序上下文之外使用 Spring AOP，但如果将两者结合，可以将拦截器和切入点都当作组件被 IoC 容器进行管理和装配。

在 Spring 中有 3 种方式去实现 AOP 的拦截功能。

(1) 使用 ProxyFactoryBean 和对应的接口实现 AOP。

(2) 使用 XML 配置 AOP。

(3) 使用 AspectJ 定义切面。

在 Spring AOP 的拦截方法中，真正常用的是用 AspectJ 方式来定义切入点和通知，这种方式下 Spring 支持基于注解的"零配置"方式，使用@AspectJ 注解进行开发，这也是开发中最常用的开发模式，有时候 XML 配置也有一定的辅助作用。本书中重点介绍使用@AspectJ 注解的方式。

AspectJ 是一个基于 Java 语言的 AOP 框架，提供了强大的 AOP 功能，Spring 的 AOP 与 AspectJ 进行了完美的集成，在使用它之前，我们需要导入 AspectJ 的依赖坐标，如下所示。

```
<!--spring-aop 依赖-->
<dependency>
```

```
        <groupId>org.aspectj</groupId>
        <artifactId>aspectjweaver</artifactId>
        <version>1.9.6</version>
    </dependency>
```

## 7.3.1　选择切点

在前面我们已经了解到切点就是需要增强处理的连接点，也就是某个类中需要增强处理的方法。先创建一个接口——UserService。代码如下所示。

```
public interface UserService {
    /**
     * 批量删除用户
     *
     * @param idList 用户编号的列表
     */
    void batchDelete(List<Integer> idList);

    /**
     * 查询用户
     *
     * @param id
     */
    void findUserById(Integer id);
}
```

接口的实现类 UserServiceImpl，代码如下所示。

```
@Service
public class UserServiceImpl implements UserService {
    @Override
    public void batchDelete(List<Integer> idList) {
        System.out.println("正在进行批量删除");
// int i = 1 / 0;
    }

    @Override
    public void findUserById(Integer id) {
        System.out.println("正在进行查询");
    }
}
```

在这个类中，有两个方法，后面只对 batchDelete 方法进行增强处理，那么这个连接点就是切点。

## 7.3.2 创建切面

在 Spring 中只要使用@Aspect 注解一个类，Spring IoC 容器就会认为这是一个切面。代码如下所示。

```
package com.bailiban.aspect;

import org.aspectj.lang.annotation.*;
import org.springframework.stereotype.Component;
/**
 * @author zhulang
 * @Classname LogAspect
 * @Description 日志切面类
 * @Date 2020/7/29 11:51
 */
@Aspect
@Component
public class LogAspect {
    @Before("execution(* com.bailiban.service.impl.UserServiceImpl.batchDelete(..))")
    public void before() {
        System.out.println("before......");
    }

    @After("execution(* com.bailiban.service.impl.UserServiceImpl.batchDelete(..))")
    public void after() {
        System.out.println("after......");
    }

    @AfterReturning("execution(* com.bailiban.service.impl.UserServiceImpl.batchDelete(..))")
    public void afterReturning() {
        System.out.println("afterReturning......");
    }

    @AfterThrowing("execution(* com.bailiban.service.impl.UserServiceImpl.batchDelete(..))")
    public void afterThrowing() {
        System.out.println("afterThrowing......");
    }
}
```

代码中@Before、@After、@AfterReturning 和@AfterThrowing 都属于 AspectJ 的注解，其释义如表 7-1 所示。

表 7-1　Spring 中的 AspectJ 注解

| 注解 | 通知 | 备注 |
|---|---|---|
| @Before | 在被代理对象的方法前先调用 | 前置通知 |
| @Around | 将被代理对象的方法封装起来，并用环绕通知取代它 | 环绕通知，它将覆盖原有方法。它可以改变执行目标方法的参数值，也可以改变执行目标方法之后的返回值 |
| @After | 在被代理对象的方法后调用 | 后置通知 |
| @AfterReturning | 在被代理对象的方法正常返回后调用 | 返回通知，要求被代理对象的方法执行过程中没有发生异常 |
| @AfterThrowing | 在被代理对象的方法抛出异常后调用 | 异常通知，要求被代理对象的方法执行过程中产生异常 |

## 7.3.3　切入点表达式

在 LogAspect 切面中，我们使用切入点表达式来告知 Spring 要对哪些方法进行拦截增强处理。表达式的代码部分如下所示。

```
execution(* com.bailiban.service.impl.UserServiceImpl.batchDelete(..))
```

关于此表达式的释义如下。

- execution：切入点指示符，代表执行方法的时候会触发。
- *：代表任意返回类型的方法。
- com.bailiban.service.impl.UserServiceImpl：代表类的全限定名。
- batchDelete：被拦截增强的方法名称。
- (..)：方法中的任意参数。

这里有常使用的 execution 表达式，其中 execution 就是一个切入点指示符。Spring AOP 仅支持部分 AspectJ 的切入点指示符，但 Spring AOP 还额外扩展了一个 Bean( )的指示符，使得我们可以根据 bean id 或者名称去定义对应的 Bean。Spring AOP 所支持 AspectJ 的指示符如表 7-2 所示。

表 7-2　AspectJ 的指示符

| AspectJ 指示符 | 描述 |
|---|---|
| args() | 用于匹配当前执行的方法传入的参数为指定类型的执行方法 |
| @args() | 用于匹配当前执行的方法传入的参数持有指定注解的执行方法 |

(续表)

| AspectJ 指示符 | 描述 |
| --- | --- |
| execution() | 用于匹配连接点的执行方法，最常用 |
| this() | 用于匹配当前 AOP 代理对象类型的执行方法 |
| target() | 用于匹配当前目标对象类型的执行方法 |
| @target() | 用于匹配当前目标对象类型的执行方法，其中目标对象持有指定的注解 |
| within() | 用于匹配指定类型内的方法执行 |
| @within() | 用于匹配所有持有指定注解类型内的方法 |
| @annotation | 用于匹配当前执行方法持有指定注解的方法 |

在 7.3.2 节中，我们在多个方法上都使用了 execution 定义的正则表达式，且代码是重复的，此时可以使用一个注解@Pointcut 来简化代码。代码如下所示。

```
@Aspect
@Component
public class LogAspect {
    @Pointcut("execution(* com.bailiban.service.impl.UserServiceImpl.batchDelete(..))")
    public void logPrint() {

    }

    @Before("logPrint()")
    public void before() {
        System.out.println("before......");
    }

    @After("logPrint()")
    public void after() {
        System.out.println("after......");
    }

    @AfterReturning("logPrint()")
    public void afterReturning() {
        System.out.println("afterReturning......");
    }

    @AfterThrowing("logPrint()")
    public void afterThrowing() {
        System.out.println("afterThrowing......");
    }
}
```

### 7.3.4 测试 AOP

上面完成了切面的各个通知和连接点的规则，下面对其进行测试，和上一单元一样，也可以使用 Java 类配置。代码如下。

```
package com.bailiban;

import org.springframework.context.annotation.ComponentScan;
import org.springframework.context.annotation.Configuration;
import org.springframework.context.annotation.EnableAspectJAutoProxy;

/**
 * @author zhulang
 * @Classname AopConfig
 * @Description 配置类
 * @Date 2020/7/29 14:45
 */
@Configuration
@EnableAspectJAutoProxy
@ComponentScan
public class AopConfig {
}
```

配置类中加粗部分的注解，代表启用 AspectJ 框架的自动代理，这个时候 Spring 才会生成动态代理对象，进而可以使用 AOP。

如果不是有 Java 类配置方式，也可以使用 XML 的方式，配置的代码 beans.xml 如下所示。

```
<?xml version="1.0" encoding="UTF-8"?>
<beans xmlns="http://www.springframework.org/schema/beans"
        xmlns:xsi="http://www.w3.org/2001/XMLSchema-instance"
        xmlns:aop="http://www.springframework.org/schema/aop"
        xmlns:context="http://www.springframework.org/schema/context"
        xsi:schemaLocation="http://www.springframework.org/schema/beans
                            http://www.springframework.org/schema/beans/spring-beans.xsd
                            http://www.springframework.org/schema/aop
                            http://www.springframework.org/schema/aop/spring-aop.xsd
                            http://www.springframework.org/schema/context
                            http://www.springframework.org/schema/context/spring-context.xsd">
    <!-- 开启 aop 自动代理-->
    <aop:aspectj-autoproxy/>
    <!-- 开启注解自动扫描-->
    <context:component-scan base-package="com.bailiban"/>
</beans>
```

测试代码如下所示：

```
package com.bailiban;

import com.bailiban.service.UserService;
import org.junit.Test;
import org.springframework.context.ApplicationContext;
import org.springframework.context.annotation.AnnotationConfigApplicationContext;
import org.springframework.context.support.ClassPathXmlApplicationContext;

/**
 * @author zhulang
 * @Classname AopTest
 * @Description aop 测试类
 * @Date 2020/7/29 14:48
 */
public class AopTest {
    /**
     * 方式一：Java 类注解配置
     */
    @Test
    public void aspectJ() {
        ApplicationContext context = new AnnotationConfigApplicationContext(AopConfig.class);
        UserService userService = context.getBean("userServiceImpl", UserService.class);
        userService.batchDelete(null);
    }

    /**
     * 方式二：XML 配置
     */
    @Test
    public void aspectJ2() {
        ApplicationContext context = new ClassPathXmlApplicationContext("beans.xml");
        UserService userService = context.getBean("userServiceImpl", UserService.class);
        userService.batchDelete(null);
    }
}
```

运行结果如图 7-5 所示。

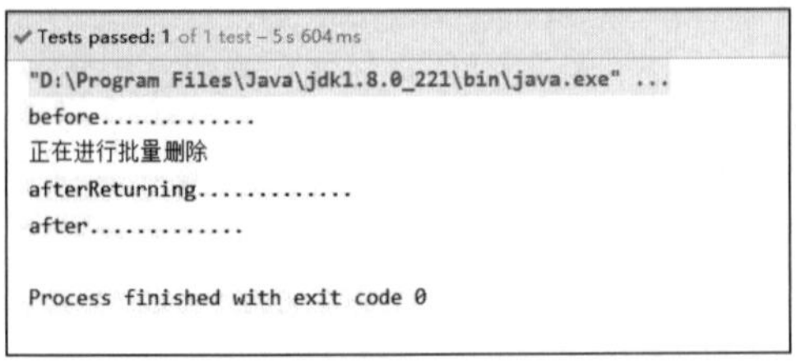

```
✔ Tests passed: 1 of 1 test – 5 s 604 ms
"D:\Program Files\Java\jdk1.8.0_221\bin\java.exe" ...
before.............
正在进行批量删除
afterReturning.............
after.............

Process finished with exit code 0
```

图 7-5

通过测试可以发现，切面的通知已经通过 AOP 植入到指定的业务中了，当切入点中有异常发生时，输出的结果又会和上面的结果不一样，大家可以自己进行测试练习。

## 7.3.5　环绕通知

环绕通知是 Spring AOP 中最强大的通知，它可以同时实现前置通知和后置通知。它的功能虽然强大，但通常需要在线程安全的环境下使用。因此如果使用普通的增强处理就能解决的问题，就没有必要使用环绕通知处理了。如果需要目标方法执行之前和之后能共享某种状态数据，则考虑使用环绕通知处理；尤其是需要改变目标方法的返回值时，就只能使用环绕通知处理了。下面在 LogAspect 类中添加如下代码。

```
@Around("logPrint()")
public void around(ProceedingJoinPoint jp){
    System.out.println("around before....");
    try {
        jp.proceed();
    } catch (Throwable throwable) {
        throwable.printStackTrace();
    }
    System.out.println("around after");
}
```

再次执行测试代码，结果如图 7-6 所示。

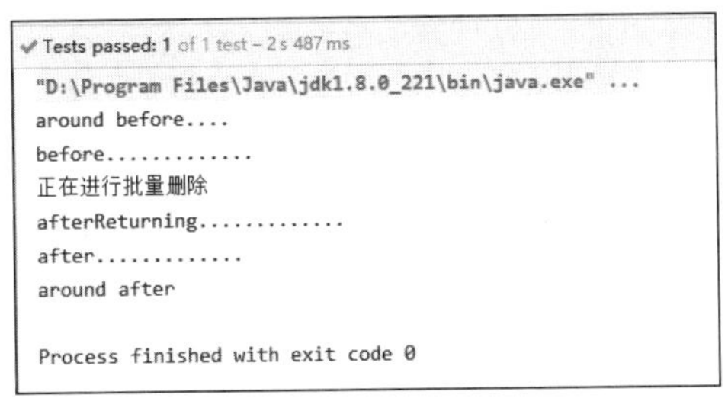


图 7-6

在上面的代码中我们传入了一个指定的参数 ProceedingJoinPoint，它是 JoinPoint 类型的子类。该参数代表了植入增强处理的连接点。里面包含了如下几个常用的方法。

- Object[] getArgs()：返回执行目标方法时的参数。
- Signature getSignature()：返回被增强的方法的相关信息。
- Object getTarget()：返回被植入增强处理的目标对象。
- Object getThis()：返回 AOP 框架为目标对象生成的代理对象。

通过这些方法就可以访问到目标方法的相关信息。

### 7.3.6 参数传递

在我们定义的切面中，只要是符合条件的连接点都会执行通知的方法，有时候可能需要向通知中传递参数，此时就可以使用如下方式实现。接下来修改 LogAspect 中的 before 方法作为演示，代码如下所示。

```
@Before("execution(* com.bailiban.service.impl.UserServiceImpl.batchDelete(..))&& args(ids)")
public void before(List<Integer> ids) {
        System.out.println("before......");
        for (Integer id : ids) {
                System.out.println("id:" + id);
        }
}
```

在上面的代码中，我们给原先的切入点表达式增加了&&args(ids)部分，当然这里只传递了一个参数，也可以传递多个参数，通过逗号隔开，那么就要求目标方法中也必须有对应参数个数和相同类型的参数。例如，修改目标方法 batchDelete 方法的参数，代码如下。

```
void batchDelete(List<Integer> idList, String name, String address);
```

然后再修改切面中的 before 方法，代码如下。

```
@Before("execution(* com.bailiban.service.impl.UserServiceImpl.batchDelete(..))&&
            args(ids,name,address)")
public void before(List<Integer> ids, String name,String address) {
        System.out.println("before......");
        for (Integer id : ids) {
                System.out.println("id:" + id);
        }
        System.out.println("name:" + name);
        System.out.println("address:" + address);
}
```

最后，在测试方法中进行测试。代码如下所示。

```
@Test
public void aspectJ2() {
        ApplicationContext context = new ClassPathXmlApplicationContext("beans.xml");
        UserService userService = context.getBean("userServiceImpl", UserService.class);
        List<Integer> list = new ArrayList<>();
        list.add(1);
        list.add(2);
        userService.batchDelete(list, "百里半", "武汉");
}
```

控制台的输出结果如图 7-7 所示。

```
✓ Tests passed: 1 of 1 test – 2 s 795 ms
"D:\Program Files\Java\jdk1.8.0_221\bin\java.exe" ...
around before....
before.............
id:1
id:2
name:百里半
address:武汉
正在进行批量删除
afterReturning.............
after.............
around after

Process finished with exit code 0
```

图 7-7

## 单元小结

- Spring 两大核心内容分别是 AOP 和 IoC。
- AOP 面向切面编程能够更好地实现关注点分离，多用于异常处理、安全、日志、事务处理切面。
- Advice 的中文术语有多种不同的叫法，如“通知”“处理”“增强处理”等。
- Spring AOP 默认使用 JDK 动态代理。
- Spring 并不支持所有的 AspectJ 切入点指示符，如果使用了不支持的指示符，将会抛出 IllegalArgumentException 异常。
- Around 环绕通知近似等于 Before 和 AfterReturning 通知的总和，功能比较强大。
- Weaving(织入)指的是将通知添加到目标对象中，并创建一个被增强的对象(AOP 代理)的过程。织入有两种实现方式：编译时增强(如 AspectJ)和运行时增强(如 Spring AOP)。

## 单元自测

1. Spring AOP 常用的术语有(　　)。

   A. Aspect　　　　B. Joinpoint

   C. Pointcut　　　　D. Weaving

2. 以下关于 Spring AOP 的介绍，错误的是(　　)。

   A. AOP 的全称是 Aspect-Oriented Programming，即面向切面编程(也称面向方面编程)

   B. AOP 采取横向抽取机制，将分散在各个方法中的重复代码提取出来，这种采用横向抽取机制的方式，采用 OOP 思想是无法办到的

C. AOP是一种新的编程思想，采取横向抽取机制，是OOP的升级替代品

D. 目前最流行的AOP框架有两个，分别为Spring AOP和AspectJ

3. 下列有关JDK动态代理的描述，正确的有(　　)。

A. JDK动态代理是通过java.lang.reflect.Proxy 类来实现的

B. 对于使用业务接口的类，Spring默认会使用JDK动态代理来实现AOP

C. Spring中的AOP代理，可以是JDK动态代理，也可以是CGLIB代理

D. 使用JDK动态代理的对象不必实现接口

4. 以下有关CGLIB代理的相关说法，错误的是(　　)。

A. CGLIB代理的使用非常简单，但它还有一定的局限性——使用动态代理的对象必须实现一个或多个接口

B. 如果要对没有实现接口的类进行代理，那么可以使用JDK代理

C. CGLIB是一个高性能开源的代码生成包，在使用时需要另外导入CGLIB所需要的包

D. Spring中的AOP代理，可以是JDK动态代理，也可以是CGLIB代理

5. 在Spring AOP中，如果想计算某一个方法执行的时间，可以使用Spring AOP中的(　　)来实现该功能。

A. Around通知　　B. Before通知

C. AfterReturning通知　　D. AfterThrowing通知

## 上机实战

### 上机目标

掌握Spring AOP中各种通知的使用。

### 上机练习

#### ◆第一阶段◆

#### 练习1：使用Spring AOP中的Before和After Returning通知

【问题描述】

给指定接口的所有方法添加日志跟踪功能，在进入和离开该方法时输出相关的信息。

【问题分析】

因为需要给方法调用前后有信息输出，可以使用 Around 通知，但也可以组合 Before 和 After Returning 通知达到同样的效果。

【参考步骤】

参考本章示例的代码，自行完成。

## ◆第二阶段◆

### 练习 2：给通知传递参数

【问题描述】

结合本章节参数传递，修改练习 1，进行传参测试。

【问题分析】

切入点表达式中用&& args()进行参数的传递。

# Spring和数据库编程

## 课程目标

- ❖ 了解传统 JDBC 代码的弊端
- ❖ 掌握如何使用 Spring 配置各种数据源
- ❖ 掌握 jdbcTemplate 的基础用法
- ❖ 掌握 Spring+MyBatis 项目的整合方法

## 简介

在 Java 互联网项目中，数据大部分存储在数据库和 NoSQL 工具中，本章将接触到数据库的编程。Spring 为开发者提供了 JDBC 模板模式，即它自身的 jdbcTemplate，它可以简化许多代码的编程，但是在实际的工作中 jdbcTemplate 并不常用。

在前面我们已经学过持久层框架——MyBatis 框架。Spring 并不代替已有框架的功能，而是以提供模板的形式给予支持。例如 Spring 对 Hibernate 框架提供了 HibernateTemplate 模板，虽然官方没有提供对 MyBatis 框架模板的支持，但是 MyBatis 社区开发了接入 Spring 的开发包，提供了 SqlSessionTemplate 模板代码。本章将讨论 Spring 和 MyBatis 框架的集成。

# 8.1 传统 JDBC 代码的弊端

为了跟 Spring 的数据库编程进行比较，我们先看一下传统 JDBC 的编码方式，代码如下所示。

```
package com.bailiban.dao.impl;

import com.bailiban.dao.RoleDao;
import com.bailiban.entity.Role;

import java.sql.*;

/**
 * @author zhulang
 * @Classname RoleDaoImpl
 * @Description TODO
 * @Date 2020/7/31 14:42
 */
public class RoleDaoImpl implements RoleDao {
    @Override
    public Role getRole(Integer id) {
        Role role = null;
// 声明 JDBC 变量
        Connection connection = null;
        PreparedStatement preparedStatement = null;
        ResultSet resultSet = null;
        try {
// 注册驱动程序
```

```
                Class.forName("com.mysql.cj.jdbc.Driver");
                connection = DriverManager.getConnection("jdbc:mysql://localhost:3306/
                mybatis?server Timezone=GMT%2b8&useSSL=false&useUnicode=
                true&characterEncoding=utf-8", "root", "123456");
                String sql = "select * from tab_role where id=?";
                preparedStatement = connection.prepareStatement(sql);
                preparedStatement.setInt(1, id);
                resultSet = preparedStatement.executeQuery();
                while (resultSet.next()) {
                    role = new Role();
                    role.setId(1);
                    role.setRoleName(resultSet.getString("role_name"));
                    role.setRoleDesp(resultSet.getString("role_desp"));
                    role.setEnable(resultSet.getInt("enable"));
                }
            } catch (ClassNotFoundException e) {
                e.printStackTrace();
            } catch (SQLException e) {
                e.printStackTrace();
            } finally {
// 关闭数据库连接资源
                try {
                    if (resultSet != null) {
                        resultSet.close();
                    }
                    if (preparedStatement != null) {
                        preparedStatement.close();
                    }
                    if (connection != null) {
                        connection.close();
                    }
                } catch (SQLException e) {
                    e.printStackTrace();
                }
            }
            return role;
        }
    }
```

上面代码中的 Role 实体类是我们在单元四中使用过的一个实体类，这里就不再展示代码。从代码中可以看出，我们进行的是根据 id 查询 Role 角色信息，业务功能十分简单，但是其代码却非常繁琐，并且最后需要关闭数据库资源，导致要写很多 try...catch...finally...等语句。这样代码的可读性和可维护性就比较差。有了 Spring 提供的 jdbcTemplate 模板，就能轻松解决上面这些问题。

# 8.2 配置数据库资源

在 Spring 中配置数据库资源并不难，大部分情况下我们都会配置成数据库连接池，这样可以节省频繁创建和销毁数据库的资源开销。在配置的时候，可以选择使用 Spring 内部提供的类，也可以使用第三方数据库连接池提供的类。下面以 XML 配置方式进行讲解。

## 8.2.1 使用 Spring-JDBC 进行数据库配置

Spring 内部提供了数据库配置类，此时需要在 pom 文件中添加 Spring-JDBC 的依赖坐标。如下所示。

```
<dependency>
    <groupId>org.springframework</groupId>
    <artifactId>spring-jdbc</artifactId>
    <version>5.2.7.RELEASE</version>
</dependency>
```

beans.xml 配置文件内容如下：

```
<?xml version="1.0" encoding="UTF-8"?>
<beans xmlns="http://www.springframework.org/schema/beans"
        xmlns:xsi="http://www.w3.org/2001/XMLSchema-instance"
        xmlns:context="http://www.springframework.org/schema/context"
        xsi:schemaLocation="http://www.springframework.org/schema/beans
                        http://www.springframework.org/schema/beans/spring-beans.xsd
                        http://www.springframework.org/schema/context
                        http://www.springframework.org/schema/context/spring-context.xsd">
    <!--读取 properties 文件-->
    <context:property-placeholder location="classpath:db.properties"/>
    <!--配置数据源-->
    <bean id="dataSource" class="org.springframework.jdbc.datasource.DriverManagerDataSource">
        <property name="driverClassName" value="${jdbc.driver}"/>
        <property name="url" value="${jdbc.url}"/>
        <property name="username" value="${jdbc.username}"/>
        <property name="password" value="${jdbc.password}"/>
    </bean>
    <!-- 开启注解自动扫描-->
    <context:component-scan base-package="com.bailiban"/>
</beans>
```

其中 db.properties 是数据库连接信息，我们使用的是之前在本书单元二中使用的数据库，内容如下。

```
jdbc.driver=com.mysql.cj.jdbc.Driver
jdbc.url=jdbc:mysql://localhost:3306/mybatis-demo3?serverTimezone=GMT%2b8&useSSL
        =false&useUnicode=true&characterEncoding=utf-8
jdbc.username=root
jdbc.password=123456
```

这样很容易就配置好了一个简单的数据源。需要注意的是，Spring 提供的 DriverManagerDataSource 并不是一个数据库连接池，每次创建连接时都会重新创建一个连接。因此并不推荐使用这种方式。更多的时候，会采用第三方的数据库连接。

## 8.2.2 使用第三方数据库连接池

上面我们提到使用 Spring 内部的连接无法应用连接池，因此使用第三方数据库连接池，例如 alibaba druid 连接池，然后在 Spring 中简单配置，就能使用了。

代码如下所示。

```
<?xml version="1.0" encoding="UTF-8"?>
<beans xmlns="http://www.springframework.org/schema/beans"
       xmlns:xsi="http://www.w3.org/2001/XMLSchema-instance"
       xmlns:context="http://www.springframework.org/schema/context"
       xsi:schemaLocation="http://www.springframework.org/schema/beans
                           http://www.springframework.org/schema/beans/spring-beans.xsd
                           http://www.springframework.org/schema/context
                           http://www.springframework.org/schema/context/spring-context.xsd">
    <!--读取 properties 文件-->
    <context:property-placeholder location="classpath:db.properties"/>
    <!--配置数据源-->
    <bean id="dataSource" class="com.alibaba.druid.pool.DruidDataSource">
        <property name="driverClassName" value="${jdbc.driver}"/>
        <property name="url" value="${jdbc.url}"/>
        <property name="username" value="${jdbc.username}"/>
        <property name="password" value="${jdbc.password}"/>
        <!--配置初始化大小、最小、最大-->
        <property name="initialSize" value="2"/>
        <property name="minIdle" value="1"/>
        <property name="maxActive" value="10"/>
        <!--配置获取连接等待超时的时间-->
        <property name="maxWait" value="60000"/>

    </bean>
    <!-- 开启注解自动扫描-->
    <context:component-scan base-package="com.bailiban"/>
</beans>
```

这样就配置好了一个 druid 的数据库连接池。

# 8.3 jdbcTemplate 的使用

前面提到过使用传统 JDBC 有一些弊端，Spring 内部提供了 jdbcTemplate 模板来简化开发者的工作量。

## 8.3.1 jdbcTemplate 的配置

使用 jdbcTemplate 需要为其注入 dataSource 数据源，其 XML 配置如下所示。

```
<?xml version="1.0" encoding="UTF-8"?>
<beans xmlns="http://www.springframework.org/schema/beans"
        xmlns:xsi="http://www.w3.org/2001/XMLSchema-instance"
        xmlns:context="http://www.springframework.org/schema/context"
        xsi:schemaLocation="http://www.springframework.org/schema/beans
                            http://www.springframework.org/schema/beans/spring-beans.xsd
                            http://www.springframework.org/schema/context
                            http://www.springframework.org/schema/context/spring-context.xsd">
    <!--读取 properties 文件-->
    <context:property-placeholder location="classpath:db.properties"/>
    <!--配置数据源-->
    <bean id="dataSource" class="com.alibaba.druid.pool.DruidDataSource">
        <property name="driverClassName" value="${jdbc.driver}"/>
        <property name="url" value="${jdbc.url}"/>
        <property name="username" value="${jdbc.username}"/>
        <property name="password" value="${jdbc.password}"/>
        <!--配置初始化大小、最小、最大-->
        <property name="initialSize" value="2"/>
        <property name="minIdle" value="1"/>
        <property name="maxActive" value="10"/>
        <!--配置获取连接等待超时的时间-->
        <property name="maxWait" value="60000"/>
    </bean>
    <bean id="jdbcTemplate" class="org.springframework.jdbc.core.JdbcTemplate">
        <property name="dataSource" ref="dataSource"/>
    </bean>
    <!--开启注解自动扫描-->
    <context:component-scan base-package="com.bailiban"/>
</beans>
```

配置好 dataSource 和 jdbcTemplate，就可以操作 jdbcTemplate 了。

## 8.3.2 jdbcTemplate 的增、删、查、改

完成 jdbcTemplate 的配置后，就可以使用 jdbcTemplate 来完成增、删、查、改操作。示例代码如下所示。

```
package com.bailiban.dao.impl;

import com.bailiban.dao.RoleDao;
import com.bailiban.entity.Role;
import org.springframework.beans.factory.annotation.Autowired;
import org.springframework.jdbc.core.JdbcTemplate;
import org.springframework.jdbc.core.RowMapper;
import org.springframework.stereotype.Repository;

/**
 * @author zhulang
 * @Classname RoleDaoImpl2
 * @Description roleDao 持久层
 * @Date 2020/8/3 10:21
 */
@Repository
public class RoleDaoImpl2 implements RoleDao {
    @Autowired
    private JdbcTemplate jdbcTemplate;

    /**
     * 根据 id 查询 role
     *
     * @param id
     * @return
     */
    @Override
    public Role getRole(Integer id) {
        String sql = "select * from tab_role where id=?";
        Role result = jdbcTemplate.queryForObject(sql, (RowMapper<Role>) (resultSet, i) -> {
            Role role = new Role();
            role.setId(id);
            role.setRoleName(resultSet.getString("role_name"));
            role.setRoleDesp(resultSet.getString("role_desp"));
            role.setEnable(resultSet.getInt("enable"));
            return role;
        }, id);
        return result;
    }

    /**
```

```
     * 插入 role
     *
     * @param role
     * @return
     */
    @Override
    public int insertRole(Role role) {
        String sql = "insert into  tab_role(role_name,role_desp,enable) values(?,?,?) ";
        Object[] params = new Object[]{role.getRoleName(), role.getRoleDesp(), role.getEnable()};
        return jdbcTemplate.update(sql, params);
    }

    /**
     * 根据 id 删除 role
     *
     * @param id
     * @return
     */
    @Override
    public int deleteRole(Integer id) {
        String sql = "delete from tab_role where id=? ";
        return jdbcTemplate.update(sql, id);
    }

    /**
     * 更新 role
     *
     * @param role
     * @return
     */
    @Override
    public int updateRole(Role role) {
        String sql = "update   tab_role set role_name=?,role_desp=?,enable=? where id=? ";
        Object[] params = new Object[]{role.getRoleName(), role.getRoleDesp(), role.getEnable(),
                                       role.getId()};
        return jdbcTemplate.update(sql, params);
    }
}
```

测试方法如下：

```
@Test
public void jdbctemplate() {
    ApplicationContext context = new ClassPathXmlApplicationContext("beans.xml");
    RoleDao roleDao = context.getBean("roleDaoImpl2", RoleDao.class);
    System.out.println("-------查询 id=3 的 role--------");
    Role role = roleDao.getRole(3);
    System.out.println(role);
```

```
        System.out.println("-------插入 role--------");
        Role role1 = new Role();
        role1.setRoleName("研发组长");
        role1.setRoleDesp("负责小组研发项目");
        role1.setEnable(1);
        int iinsert = roleDao.insertRole(role1);
        System.out.println(iinsert);
        System.out.println("-------更新 role--------");
        role.setRoleDesp("负责分管华中区域");
        int iupdate = roleDao.updateRole(role);
        System.out.println(iupdate);
        System.out.println("-------删除 id=13 的 role--------");
        int idelete = roleDao.deleteRole(13);
        System.out.println(idelete);
    }
```

测试运行的结果如图 8-1 所示。

```
✔ Tests passed: 1 of 1 test – 3 s 336 ms
"D:\Program Files\Java\jdk1.8.0_221\bin\java.exe" ...
-------查询id=3的role--------
八月 05, 2020 3:16:58 下午 com.alibaba.druid.support.logging.JakartaCommonsLoggingImpl info
信息: {dataSource-1} inited
Role(id=3, roleName=副总经理, roleDesp=分管其中一个片区, enable=1, emps=null, permissions=null)
-------插入role--------
1
-------更新role--------
1
-------删除id=13的role--------
1

Process finished with exit code 0
```

图 8-1

## 8.4 Spring+MyBatis 的整合

尽管在使用 Spring 时可以使用其提供的 jdbcTemplate 来完成持久化的操作，但是对于一些复杂的 SQL，处理起来还是比较繁琐。而目前持久层主流框架之一的 MyBatis 就可以更好地为开发者提供持久层服务。在本书的最开始几个单元中，已经讲解了 MyBatis 的相关知识，下面将其和 Spring 进行整合。

在 Spring 中使用 MyBatis 需要依赖其相关 jar 包，在 pom 中添加如下依赖坐标。

```
<dependency>
    <groupId>org.mybatis</groupId>
    <artifactId>mybatis</artifactId>
    <version>3.5.5</version>
```

```
    </dependency>
    <dependency>
        <groupId>org.mybatis</groupId>
        <artifactId>mybatis-spring</artifactId>
        <version>2.0.3</version>
    </dependency>
```

## 8.4.1 MyBatis 的配置

实际上 MyBatis 的相关配置可以合并到 Spring 中统一进行，但是为了提高可读性和便于维护，建议分开配置。其配置文件 mybatis-config.xml 的代码如下所示。

```
<?xml version="1.0" encoding="UTF-8"?>
<!DOCTYPE configuration PUBLIC "-//mybatis.org//DTD Config 3.0//EN"
"http://mybatis.org/dtd/mybatis-3-config.dtd">
<configuration><!--配置-->
    <settings>
        <!--指定 MyBatis 使用 log4j -->
        <setting name="logImpl" value="LOG4J"/>
        <!--开启延迟加载功能-->
        <setting name="lazyLoadingEnabled" value="true"/>
        <!--不进行完整对象的加载，属性按需加载-->
        <setting name="aggressiveLazyLoading" value="false"/>
    </settings>
    <typeAliases>
        <!--类型别名-->
        <package name="com.bailiban.entity"/>
    </typeAliases>
    <mappers>
        <mapper resource="mapper/RoleMapper.xml"/>
    </mappers>
</configuration>
```

其中 RoleMapper 接口的映射文件放在 resources 资源目录下的 mapper 目录下，其内容如下。

```
<?xml version="1.0" encoding="UTF-8"?>
<!DOCTYPE mapper PUBLIC "-//mybatis.org//DTD Mapper 3.0//EN"
"http://mybatis.org/dtd/mybatis-3-mapper.dtd">
<mapper namespace="com.bailiban.mapper.RoleMapper">

    <resultMap type="com.bailiban.entity.Role" id="RoleMap">
    <result property="id" column="id" jdbcType="INTEGER"/>
    <result property="roleName" column="role_name" jdbcType="VARCHAR"/>
    <result property="roleDesp" column="role_desp" jdbcType="VARCHAR"/>
    <result property="enable" column="enable" jdbcType="INTEGER"/>
```

```
</resultMap>

<!--查询单个-->
<select id="queryById" resultMap="RoleMap">
    select id,
          role_name,
          role_desp,
          enable
    from tab_role
    where id = #{id}
</select>

<!--查询指定行数据-->
<select id="queryAllByLimit" resultMap="RoleMap">
    select id,
           role_name,
           role_desp,
           enable
    from tab_role
    limit #{offset}, #{limit}
</select>

<!--通过实体作为筛选条件查询-->
<select id="queryAll" resultMap="RoleMap">
    select
          id, role_name, role_desp, enable
    from tab_role
    <where>
          <if test="id != null">
                and id = #{id}
          </if>
          <if test="roleName != null and roleName != ''">
                and role_name = #{roleName}
          </if>
          <if test="roleDesp != null and roleDesp != ''">
                and role_desp = #{roleDesp}
          </if>
          <if test="enable != null">
                and enable = #{enable}
          </if>
    </where>
</select>

<!--新增所有列-->
<insert id="insert" keyProperty="id" useGeneratedKeys="true">
    insert into tab_role(role_name, role_desp, enable)
    values (#{roleName}, #{roleDesp}, #{enable})
```

```
    </insert>

    <!--通过主键修改数据-->
    <update id="update">
        update tab_role
        <set>
            <if test="roleName != null and roleName != ''">
                role_name = #{roleName},
            </if>
            <if test="roleDesp != null and roleDesp != ''">
                role_desp = #{roleDesp},
            </if>
            <if test="enable != null">
                enable = #{enable},
            </if>
        </set>
        where id = #{id}
    </update>

    <!--通过主键删除-->
    <delete id="deleteById">
        delete
        from tab_role
        where id = #{id}
    </delete>

</mapper>
```

对应的接口 RoleMapper 代码如下。

```
package com.bailiban.mapper;

import com.bailiban.entity.Role;
import org.apache.ibatis.annotations.Param;

import java.util.List;

/**
 * (Role)表数据库访问层
 *
 * @author zhulang
 * @since 2020-08-06 10:38:52
 */
public interface RoleMapper {

    /**
     * 通过 ID 查询单条数据
     *
```

```
     * @param id  主键
     * @return  实例对象
     */
    Role queryById(Integer id);

    /**
     *  查询指定行数据
     *
     * @param offset  查询起始位置
     * @param limit  查询条数
     * @return  对象列表
     */
    List<Role> queryAllByLimit(@Param("offset") int offset, @Param("limit") int limit);

    /**
     *  通过实体作为筛选条件查询
     *
     * @param tabRole  实例对象
     * @return  对象列表
     */
    List<Role> queryAll(Role tabRole);

    /**
     *  新增数据
     *
     * @param tabRole  实例对象
     * @return  影响行数
     */
    int insert(Role tabRole);

    /**
     *  修改数据
     *
     * @param tabRole  实例对象
     * @return  影响行数
     */
    int update(Role tabRole);

    /**
     *  通过主键删除数据
     *
     * @param id  主键
     * @return  影响行数
     */
    int deleteById(Integer id);

}
```

## 8.4.2　Spring+MyBatis 的配置

在 resources 下新建 spring-mybatis.xml 文件，其内容如下。

```
<?xml version="1.0" encoding="UTF-8"?>
<beans xmlns="http://www.springframework.org/schema/beans"
        xmlns:xsi="http://www.w3.org/2001/XMLSchema-instance"
        xmlns:context="http://www.springframework.org/schema/context"
        xsi:schemaLocation="http://www.springframework.org/schema/beans
                            http://www.springframework.org/schema/beans/spring-beans.xsd
                            http://www.springframework.org/schema/context
                            http://www.springframework.org/schema/context/spring-context.xsd">
    <!--读取 properties 文件-->
    <context:property-placeholder location="classpath:db.properties"/>
    <!--开启注解自动扫描-->
    <context:component-scan base-package="com.bailiban"/>
    <!--配置数据源-->
    <bean id="dataSource" class="com.alibaba.druid.pool.DruidDataSource">
        <property name="driverClassName" value="${jdbc.driver}"/>
        <property name="url" value="${jdbc.url}"/>
        <property name="username" value="${jdbc.username}"/>
        <property name="password" value="${jdbc.password}"/>
        <!--配置初始化大小、最小、最大-->
        <property name="initialSize" value="2"/>
        <property name="minIdle" value="1"/>
        <property name="maxActive" value="10"/>
        <!--配置获取连接等待超时的时间-->
        <property name="maxWait" value="60000"/>
    </bean>
    <!--配置 MyBatis -->
    <bean id="sqlSessionFactory" class="org.mybatis.spring.SqlSessionFactoryBean">
        <property name="dataSource" ref="dataSource"/>
        <property name="configLocation" value="classpath:mybatis-config.xml"/>
    </bean>
    <bean class="org.mybatis.spring.mapper.MapperScannerConfigurer">
        <property name="basePackage" value="com.bailiban.mapper"/>
        <property name="sqlSessionFactoryBeanName" value="sqlSessionFactory"/>
    </bean>
</beans>
```

spring-mybatis.xml 配置文件的写法有很多种，在此只展示了最常用的一种，在上面配置文件中，关于 MyBatis 配置的部分一共有两个，一个是 SqlSessionFactory，另一个是 MapperScannerConfigurer，下面分别进行解释。

- SqlSessionFactory：用于产生 SqlSession 的工厂类，只允许有一个。有一个必须属性 dataSource 来指定数据源，另外还有一个通用配置 configLocation，用来指定 MyBatis 的 xml 配置文件位置。
- MapperScannerConfigurer：查找类路径下的映射器并自动将它们创建成 MapperFactoryBean。其 basePackage 属性是要扫描的映射接口文件包路径；另一个属性 SqlSessionFactoryBeanName 用于指定 SqlSessionFactory，但是对于只有一个数据源的情况，可以省去此属性的配置，因为在创建 MapperFactoryBean 后会自动进行装配。

### 8.4.3 Spring+MyBatis 的测试

完成以上代码后，我们开始对 Spring 和 MyBatis 进行测试。其中 RoleService 代码不变，依旧是测试其增删查改，对 RoleServiceImpl 代码修改如下。

```
package com.bailiban.service.impl;

import com.bailiban.entity.Role;
import com.bailiban.mapper.RoleMapper;
import com.bailiban.service.RoleService;
import org.springframework.beans.factory.annotation.Autowired;
import org.springframework.stereotype.Service;

/**
 * @author zhulang
 * @Classname RoleServiceImpl
 * @Description role 业务层
 * @Date 2020/8/5 16:14
 */
    @Service
    public class RoleServiceImpl implements RoleService {
    @Autowired
    private RoleMapper roleMapper;

    @Override
    public Role getRole(Integer id) {

        return roleMapper.queryById(id);
    }

    @Override
    public int insertRole(Role role) {

        return roleMapper.insert(role);
```

```
    }

    @Override
    public int deleteRole(Integer id) {

        return roleMapper.deleteById(id);
    }

    @Override
    public int updateRole(Role role) {

        return roleMapper.update(role);
    }
}
```

测试类代码如下。

```
package com.bailiban;

import com.bailiban.entity.Role;
import com.bailiban.service.RoleService;
import org.junit.Test;
import org.junit.runner.RunWith;
import org.springframework.beans.factory.annotation.Autowired;
import org.springframework.test.context.ContextConfiguration;
import org.springframework.test.context.junit4.SpringJUnit4ClassRunner;

/**
 * Unit test for simple App.
 */
    @RunWith(SpringJUnit4ClassRunner.class)
    @ContextConfiguration(locations = { "classpath:spring-mybatis.xml" })
    public class AppTest {
    @Autowired
    private RoleService roleService;
    @Test
    public void testSpringMybatis() {
        System.out.println("-------查询 id=3 的 role--------");
        Role role = roleService.getRole(3);
        System.out.println(role);
        System.out.println("-------插入 role--------");
        Role role1 = new Role();
        role1.setRoleName("研发组长");
        role1.setRoleDesp("负责小组研发项目");
        role1.setEnable(1);
        int iinsert = roleService.insertRole(role1);
        System.out.println(iinsert);
        System.out.println("-------更新 role--------");
```

```
            role.setRoleDesp("负责分管湖北区域");
            int iupdate = roleService.updateRole(role);
            System.out.println(iupdate);
            System.out.println("-------删除 id=13 的 role--------");
            int idelete = roleService.deleteRole(13);
            System.out.println(idelete);
        }
}
```

这次测试类中使用了 spring-test 包，其依赖坐标如下。

```
<dependency>
    <groupId>org.springframework</groupId>
    <artifactId>spring-test</artifactId>
    <version>5.2.7.RELEASE</version>
</dependency>
```

在 AppTest 类上使用了两个注解，一个是@RunWith，另一个是@ContextConfiguration，其解释如下。

- @RunWith：就是一个运行器，需要一个 Runner 类型参数，这里传递的是 SpringJUnit4ClassRunner.class，让测试运行于 Spring 测试环境。
- @ContextConfiguration：Spring 整合 JUnit4 测试时，使用注解引入多个配置文件。属性 locations 中可以配置多个配置文件且以逗号隔开。

运行测试方法，结果如图 8-2 所示。

从图 8-2 的运行结果中可以详细查看运行日志记录，因为在 MyBatis 的配置文件中实现了 log4j，同时在 pom 文件中添加了其依赖坐标。log4j 依赖坐标如下所示。

```
<dependency>
    <groupId>log4j</groupId>
    <artifactId>log4j</artifactId>
    <version>1.2.17</version>
</dependency>
```

此外从日志中可以看到每当使用一个 RoleMapper 接口的方法，就会产生一个新的 SqlSession，运行完成后就会自动关闭。从关闭的日志中可以看到“Closing non transactional SqlSession”的字样，说明它是在一个非事务的场景下运行，所以这里只是简单地使用数据库，而并没有启动数据库事务。更加深入的部分，将在下一章中进行讲解。

```
✓ Tests passed: 1 of 1 test – 1 s 757 ms
2020-08-06 14:23:33,029 [main] DEBUG [org.apache.ibatis.io.ResolverUtil] - Checking to see if class com.bailiban.entity.Role
 matches criteria [is assignable to Object]
-------查询id=3的role--------
2020-08-06 14:23:33,683 [main] DEBUG [org.mybatis.spring.SqlSessionUtils] - Creating a new SqlSession
2020-08-06 14:23:33,699 [main] DEBUG [org.mybatis.spring.SqlSessionUtils] - SqlSession [org.apache.ibatis.session.defaults
 .DefaultSqlSession@66ea810] was not registered for synchronization because synchronization is not active
2020-08-06 14:23:34,804 [main] INFO  [com.alibaba.druid.pool.DruidDataSource] - {dataSource-1} inited
2020-08-06 14:23:34,813 [main] DEBUG [org.mybatis.spring.transaction.SpringManagedTransaction] - JDBC Connection [com.mysql
 .cj.jdbc.ConnectionImpl@524f3b3a] will not be managed by Spring
2020-08-06 14:23:34,829 [main] DEBUG [com.bailiban.mapper.RoleMapper.queryById] - ==>  Preparing: select id, role_name,
 role_desp, enable from tab_role where id = ?
2020-08-06 14:23:34,991 [main] DEBUG [com.bailiban.mapper.RoleMapper.queryById] - ==> Parameters: 3(Integer)
2020-08-06 14:23:35,107 [main] DEBUG [com.bailiban.mapper.RoleMapper.queryById] - <==      Total: 1
2020-08-06 14:23:35,110 [main] DEBUG [org.mybatis.spring.SqlSessionUtils] - Closing non transactional SqlSession [org.apache
 .ibatis.session.defaults.DefaultSqlSession@66ea810]
Role(id=3, roleName=副总经理, roleDesp=负责分管华中区域, enable=1)
-------插入role--------
2020-08-06 14:23:35,112 [main] DEBUG [org.mybatis.spring.SqlSessionUtils] - Creating a new SqlSession
2020-08-06 14:23:35,112 [main] DEBUG [org.mybatis.spring.SqlSessionUtils] - SqlSession [org.apache.ibatis.session.defaults
 .DefaultSqlSession@48e92c5c] was not registered for synchronization because synchronization is not active
2020-08-06 14:23:35,112 [main] DEBUG [org.mybatis.spring.transaction.SpringManagedTransaction] - JDBC Connection [com.mysql
 .cj.jdbc.ConnectionImpl@524f3b3a] will not be managed by Spring
2020-08-06 14:23:35,113 [main] DEBUG [com.bailiban.mapper.RoleMapper.insert] - ==>  Preparing: insert into tab_role
 (role_name, role_desp, enable) values (?, ?, ?)
2020-08-06 14:23:35,115 [main] DEBUG [com.bailiban.mapper.RoleMapper.insert] - ==> Parameters: 研发组长(String), 负责小组研发
 项目(String), 1(Integer)
2020-08-06 14:23:35,153 [main] DEBUG [com.bailiban.mapper.RoleMapper.insert] - <==    Updates: 1
2020-08-06 14:23:35,163 [main] DEBUG [org.mybatis.spring.SqlSessionUtils] - Closing non transactional SqlSession [org.apache
 .ibatis.session.defaults.DefaultSqlSession@48e92c5c]
1
-------更新role--------
2020-08-06 14:23:35,164 [main] DEBUG [org.mybatis.spring.SqlSessionUtils] - Creating a new SqlSession
2020-08-06 14:23:35,164 [main] DEBUG [org.mybatis.spring.SqlSessionUtils] - SqlSession [org.apache.ibatis.session.defaults
 .DefaultSqlSession@2a62b5bc] was not registered for synchronization because synchronization is not active
2020-08-06 14:23:35,327 [main] DEBUG [org.mybatis.spring.transaction.SpringManagedTransaction] - JDBC Connection [com.mysql
 .cj.jdbc.ConnectionImpl@524f3b3a] will not be managed by Spring
2020-08-06 14:23:35,327 [main] DEBUG [com.bailiban.mapper.RoleMapper.update] - ==>  Preparing: update tab_role SET role_name
 = ?, role_desp = ?, enable = ? where id = ?
2020-08-06 14:23:35,329 [main] DEBUG [com.bailiban.mapper.RoleMapper.update] - ==> Parameters: 副总经理(String), 负责分管湖北
 区域(String), 1(Integer), 3(Integer)
2020-08-06 14:23:35,334 [main] DEBUG [com.bailiban.mapper.RoleMapper.update] - <==    Updates: 1
2020-08-06 14:23:35,334 [main] DEBUG [org.mybatis.spring.SqlSessionUtils] - Closing non transactional SqlSession [org.apache
 .ibatis.session.defaults.DefaultSqlSession@2a62b5bc]
1
-------删除id=13的role--------
2020-08-06 14:23:35,335 [main] DEBUG [org.mybatis.spring.SqlSessionUtils] - Creating a new SqlSession
2020-08-06 14:23:35,335 [main] DEBUG [org.mybatis.spring.SqlSessionUtils] - SqlSession [org.apache.ibatis.session.defaults
 .DefaultSqlSession@db57326] was not registered for synchronization because synchronization is not active
2020-08-06 14:23:35,335 [main] DEBUG [org.mybatis.spring.transaction.SpringManagedTransaction] - JDBC Connection [com.mysql
 .cj.jdbc.ConnectionImpl@524f3b3a] will not be managed by Spring
2020-08-06 14:23:35,335 [main] DEBUG [com.bailiban.mapper.RoleMapper.deleteById] - ==>  Preparing: delete from tab_role where
 id = ?
2020-08-06 14:23:35,336 [main] DEBUG [com.bailiban.mapper.RoleMapper.deleteById] - ==> Parameters: 13(Integer)
2020-08-06 14:23:35,340 [main] DEBUG [com.bailiban.mapper.RoleMapper.deleteById] - <==    Updates: 1
2020-08-06 14:23:35,340 [main] DEBUG [org.mybatis.spring.SqlSessionUtils] - Closing non transactional SqlSession [org.apache
 .ibatis.session.defaults.DefaultSqlSession@db57326]
1
2020-08-06 14:23:35,407 [SpringContextShutdownHook] INFO  [com.alibaba.druid.pool.DruidDataSource] - {dataSource-1} closing
 ...
2020-08-06 14:23:35,476 [SpringContextShutdownHook] INFO  [com.alibaba.druid.pool.DruidDataSource] - {dataSource-1} closed

Process finished with exit code 0
```

图 8-2

## 单元小结

- Spring 提供的 DriverManagerDataSource 并没有实现数据库连接池功能。
- Spring 支持使用第三方数据库连接池。
- Spring 内部提供了 jdbcTemplate 模板来简化开发。
- spring-mybatis 配置文件中可以配置一个 SqlSessionFactoryBean，通过属性 configLocation 来引入 MyBatis 的配置。

## 单元自测

1. MapperFactoryBean是MyBatis-Spring团队提供的用于根据Mapper接口生成Mapper对象的类，该类在Spring配置文件中可以配置的参数不包括(　　)。

A. mapperInterface　　B. SqlSessionFactory

C. SqlSessionTemplate　　D. basePackage

2. MapperScannerConfigurer类在Spring配置文件中使用时，可以配置的属性及说明错误的是(　　)。

A. basePackage：指定映射接口文件所在的包路径，当需要扫描多个包时可以使用分号或逗号作为分隔符

B. annotationClass：指定要扫描的注解名称，只有被注解标识的类才会被配置为映射器

C. SqlSessionFactoryBeanName：指定在Spring中定义的SqlSessionFactory的Bean名称

D. SqlSessionTemplateBeanName：指定在Spring中定义的SqlSessionTemplate的Bean名称。如果定义此属性，则SqlSessionFactoryBeanName将起作用

3. 以下不属于MapperScannerConfigurer类，在Spring配置文件中使用时需要配置的属性的是(　　)。

A. basePackage　　B. annotationClass

C. SqlSessionFactoryBeanName　　D. mapperInterface

4. 进行MyBatis与Spring的整合配置文件编写时，不包括(　　)。

A. db.properties　　B. applicationContext.xml

C. mybatis-config.xml　　D. springmvc-config.xml

5. 要实现MyBatis与Spring的整合，所需要的JAR包包括(　　)。

A. Spring框架的相关JAR包　　B. MyBatis框架相关的JAR包

C. Spring与MyBatis整合JAR包　　D. 数据库驱动包

## 上机实战

### 上机目标

- 掌握Spring+MyBatis整合。

## 上机练习

### 练习：将本章中的示例改写为使用一个 spring-mybatis 配置文件

**【问题描述】**

本章中使用了 mybatis-config.xml 和 spring-mybatis.xml 两个配置文件，对于简单的配置内容，完全可以将 mybatis-config.xml 的配置内容整合到 spring-mybatis.xml 中。

**【问题分析】**

将 mybatis-config.xml 中的映射配置文件和别名的相关配置放到 spring-mybatis.xml 中即可。

**【参考步骤】**

spring-mybatis.xml 的配置如下。

```
<?xml version="1.0" encoding="UTF-8"?>
<beans xmlns="http://www.springframework.org/schema/beans"
        xmlns:xsi="http://www.w3.org/2001/XMLSchema-instance"
        xmlns:context="http://www.springframework.org/schema/context"
        xsi:schemaLocation="http://www.springframework.org/schema/beans
                        http://www.springframework.org/schema/beans/spring-beans.xsd
                        http://www.springframework.org/schema/context
                        http://www.springframework.org/schema/context/spring-context.xsd">
    <!--读取 properties 文件-->
    <context:property-placeholder location="classpath:db.properties"/>
    <!--开启注解自动扫描-->
    <context:component-scan base-package="com.bailiban"/>
    <!--配置数据源-->
    <bean id="dataSource" class="com.alibaba.druid.pool.DruidDataSource">
        <property name="driverClassName" value="${jdbc.driver}"/>
        <property name="url" value="${jdbc.url}"/>
        <property name="username" value="${jdbc.username}"/>
        <property name="password" value="${jdbc.password}"/>
        <!--配置初始化大小、最小、最大-->
        <property name="initialSize" value="2"/>
        <property name="minIdle" value="1"/>
        <property name="maxActive" value="10"/>
        <!--配置获取连接等待超时的时间-->
        <property name="maxWait" value="60000"/>
    </bean>
    <!--配置 MyBatis -->
    <bean id="sqlSessionFactory" class="org.mybatis.spring.SqlSessionFactoryBean">
        <property name="dataSource" ref="dataSource"/>
        <!--<property name="configLocation" value="classpath:mybatis-config.xml"/>-->
        <!--别名-->
        <property name="typeAliasesPackage" value="com.bailiban.entity"/>
```

```
            <!--映射接口配置文件-->
            <property name="mapperLocations" value="mapper/RoleMapper.xml"/>
        </bean>
        <bean class="org.mybatis.spring.mapper.MapperScannerConfigurer">
            <property name="basePackage" value="com.bailiban.mapper"/>
            <property name="sqlSessionFactoryBeanName" value="sqlSessionFactory"/>
        </bean>
    </beans>
```

单元

# 九

# Spring数据库事务管理

## 课程目标

- ❖ 掌握 Spring 数据库事务管理器的基础知识
- ❖ 掌握 Spring 数据库事务管理器提交和回滚事务的规则
- ❖ 掌握数据库的 ACID 特性
- ❖ 正确使用注解@Transactional

**简介**

数据库事务是企业应用最为重要的内容之一，上一单元中讨论过 Spring 数据库编程，本单元将讨论数据库事务管理。本单元将讲解 Spring 中最常用的事务管理注解 @Transactional，介绍事务的隔离级别等内容。

# 9.1 Spring 数据库事务管理器的设计

在 Spring 中数据库事务是通过 org.springframework.transaction.PlatformTransactionManager 进行管理的，在上一单元中，我们也了解到 jdbcTemplate 是不支持事务的，而支持事务管理器的模板是 org.springframework.transaction.support.TransactionTemplate。

Spring 具体的事务管理由 PlatformTransactionManager 的不同实现类来完成。在 Spring 容器中配置 PlatformTransactionManager Bean 时，必须针对不同的环境提供不同的实现类。本书使用的是 MyBatis 框架，用得最多的事务管理器是 org.springframework.jdbc. datasource 包下的 DataSourceTransactionManager。下面配置一个事务管理器，代码如下所示。

```
<?xml version="1.0" encoding="UTF-8"?>
<beans xmlns="http://www.springframework.org/schema/beans"
        xmlns:xsi="http://www.w3.org/2001/XMLSchema-instance"
        xmlns:context="http://www.springframework.org/schema/context"
        xmlns:aop="http://www.springframework.org/schema/aop"
        xmlns:tx="http://www.springframework.org/schema/tx"
        xsi:schemaLocation="http://www.springframework.org/schema/beans
                            http://www.springframework.org/schema/beans/spring-beans.xsd
                            http://www.springframework.org/schema/context
                            http://www.springframework.org/schema/context/spring-context.xsd
                            http://www.springframework.org/schema/tx
                            http://www.springframework.org/schema/tx/spring-tx.xsd
                            http://www.springframework.org/schema/aop
                            http://www.springframework.org/schema/aop/spring-aop.xsd">
    <!--读取 properties 文件-->
    <context:property-placeholder location="classpath:db.properties"/>
    <!--开启注解自动扫描-->
    <context:component-scan base-package="com.bailiban"/>
    <!--配置数据源-->
    <bean id="dataSource" class="com.alibaba.druid.pool.DruidDataSource">
        <property name="driverClassName" value="${jdbc.driver}"/>
        <property name="url" value="${jdbc.url}"/>
        <property name="username" value="${jdbc.username}"/>
```

```
            <property name="password" value="${jdbc.password}"/>
            <!--配置初始化大小、最小、最大-->
            <property name="initialSize" value="2"/>
            <property name="minIdle" value="1"/>
            <property name="maxActive" value="10"/>
            <!--配置获取连接等待超时的时间-->
            <property name="maxWait" value="60000"/>
        </bean>
    <!--配置事务管理器-->
        <bean id="transactionManager"
         class="org.springframework.jdbc.datasource.DataSourceTransactionManager">
            <property name="dataSource" ref="dataSource"/>
        </bean>
        <bean id="jdbcTemplate" class="org.springframework.jdbc.core.JdbcTemplate">
            <property name="dataSource" ref="dataSource"/>
        </bean>
    </beans>
```

上面配置中，我们定义了数据库连接池，使用 druid 数据源实现，使用 DataSourceTransactionManager 定义数据库事务管理器，并注入了数据库连接池。这样 Spring 就知道已将数据库事务委托给事务管理器 transactionManager 管理了。

在 Spring 中可以使用声明式事务或编程式事务，现今编程式事务几乎不用了，因为它会产生冗余，代码可读性较差，所以本书只讲解声明式事务。声明式事务又可以分为 XML 配置和注解事务，其中注解事务才是目前的主流。后面我们会进行逐一讲解。

## 9.2 声明式事务

编程式事务是一种约定型的事务，在大部分情况下，当使用数据库事务时在代码中发生了异常，需要回滚事务，而不发生异常时则提交事务，从而保证数据库数据的一致性。从这点出发，Spring 给了一个约定，如果使用的是声明式事务，那么当你的业务方法不发生异常(或者发生异常，但该异常也被配置信息允许提交事务)时，Spring 就会让事务管理器提交事务，而发生异常(并且该异常不被配置信息允许提交事务)时，则让事务管理器回滚事务。

不论使用何种持久化策略，Spring 都提供了一致的事务抽象。当采用编程式事务时，开发者使用的是 Spring 事务抽象，而无须使用任何具体的底层事务 API。当使用声明式事务时，开发者无须书写任何事务管理代码，不依赖 Spring 或者任何其他事务 API。Spring 的声明式事务无须任何额外的容器支持，Spring 容器本身管理声明式事务。使用声明式事务策略，可以让开发者更好地专注于业务逻辑的实现。

### 9.2.1 @Transactional 的使用

首先我们来看@Transactional 注解的源码，如下所示。

```
package org.springframework.transaction.annotation;

import java.lang.annotation.Documented;
import java.lang.annotation.ElementType;
import java.lang.annotation.Inherited;
import java.lang.annotation.Retention;
import java.lang.annotation.RetentionPolicy;
import java.lang.annotation.Target;
import org.springframework.core.annotation.AliasFor;

@Target({ElementType.TYPE, ElementType.METHOD})
@Retention(RetentionPolicy.RUNTIME)
@Inherited
@Documented
public @interface Transactional {
    @AliasFor("transactionManager")
    String value() default "";

    @AliasFor("value")
    String transactionManager() default "";

    Propagation propagation() default Propagation.REQUIRED;

    Isolation isolation() default Isolation.DEFAULT;

    int timeout() default -1;

    boolean readOnly() default false;

    Class<? extends Throwable>[] rollbackFor() default {};

    String[] rollbackForClassName() default {};

    Class<? extends Throwable>[] noRollbackFor() default {};

    String[] noRollbackForClassName() default {};
}
```

从以上源码中可以看出 Transactional 可配置的内容并不算多，但是都比较重要，并且该注解既可以用于修饰 Spring Bean 类，也可以修饰 Bean 类中的某个方法。

如果使用@Transactional 修饰 Bean 类，则表明这些事务设置对整个 Bean 类起作用；如果使用@Transactional 修饰 Bean 的某个方法，则表明这些事务设置只对该方法有效。

关于使用@Transactional 时可指定的属性解释如下。

- isolation：用于指定事务的隔离级别。默认为底层事务的隔离级别。
- noRollbackFor：指定遇到特定异常时强制不回滚事务。
- noRollbackForClassName：指定遇到特定的多个异常时强制不回滚事务。该属性值可以指定多个异常类名。
- propagation：指定事务传播行为。
- readOnly：指定事务是否只读。
- rollbackFor：指定遇到特定异常时强制回滚事务。
- rollbackForClassName：指定遇到特定的多个异常时强制回滚事务。该属性值可以指定多个异常类名。
- timeout：指定事务的超时时长。单位为秒，当超时时，会引发异常，默认会导致事务回滚。

使用声明式事务需要配置注解驱动，在配置文件中加上如下配置就可以使用@Transactional 配置事务了。

```
<!--根据注解生成事务代理-->
<tx:annotation-driven transaction-manager="transactionManager"/>
```

## 9.2.2 使用 XML 配置事务管理器

前面讲过了最常用的注解形式的声明式事务管理，虽然 XML 方式不常用，但是它却用到了 AOP 技术。由于 Spring 采用 AOP 的方式管理事务，因此可以在事务回滚动作中插入用户自己的动作，而不仅仅是执行系统默认的回滚。

Spring 的 XML Schema 方式提供了简洁的事务配置策略，Spring 提供了 tx:命名空间来配置事务管理，tx:命名空间下提供了<tx:advisor.../>元素来配置事务增强处理，一旦使用该元素配置了事务增强处理，就可以使用<aop:advisor.../>元素启动自动代理。

配置<tx:advice.../>元素除了需要 transaction-manager 属性指定事务管理器之外，还需要配置一个<attributes.../>子元素，该子元素里又可包含多个<method.../>子元素。

配置<method.../>子元素可以指定如下几个属性。

- name：必选属性，与该事务语义关联的方法名。该属性支持使用通配符，例如 get*。
- isolation：用于指定事务的隔离级别，属性值为枚举类型。默认值为 Isolation.DEFAULT。
- rollback-for：指定触发事务回滚的异常类(应使用全限定名)。
- no-rollback-for：指定不触发事务回滚的异常类(应使用全限定名)。

- propagation：指定事务传播行为，属性值为枚举类型。默认值为 Propagation. REQUIRED。
- read-only：指定事务是否只读。默认值为 false。
- timeout：指定事务的超时时长。单位为秒，当超时时，会引发异常，默认会导致事务回滚。默认值为-1，表示不超时。

从上面的属性来看，和使用注解@Transactional 的配置基本一致。下面我们展示 XML 方式下的事务管理配置。代码如下所示。

```
<?xml version="1.0" encoding="UTF-8"?>
<beans xmlns="http://www.springframework.org/schema/beans"
        xmlns:xsi="http://www.w3.org/2001/XMLSchema-instance"
        xmlns:context="http://www.springframework.org/schema/context"
        xmlns:aop="http://www.springframework.org/schema/aop"
        xmlns:tx="http://www.springframework.org/schema/tx"
        xsi:schemaLocation="http://www.springframework.org/schema/beans
                    http://www.springframework.org/schema/beans/spring-beans.xsd
                    http://www.springframework.org/schema/context
                    http://www.springframework.org/schema/context/spring-context.xsd
                    http://www.springframework.org/schema/tx
                    http://www.springframework.org/schema/tx/spring-tx.xsd
                    http://www.springframework.org/schema/aop
                    http://www.springframework.org/schema/aop/spring-aop.xsd">
    <!--读取 properties 文件-->
    <context:property-placeholder location="classpath:db.properties"/>
    <!--开启注解自动扫描-->
    <context:component-scan base-package="com.bailiban"/>
    <!--配置数据源-->
    <bean id="dataSource" class="com.alibaba.druid.pool.DruidDataSource">
        <property name="driverClassName" value="${jdbc.driver}"/>
        <property name="url" value="${jdbc.url}"/>
        <property name="username" value="${jdbc.username}"/>
        <property name="password" value="${jdbc.password}"/>
        <!--配置初始化大小、最小、最大-->
        <property name="initialSize" value="2"/>
        <property name="minIdle" value="1"/>
        <property name="maxActive" value="10"/>
        <!--配置获取连接等待超时的时间-->
        <property name="maxWait" value="60000"/>
    </bean>
    <!--配置事务管理器-->
    <bean id="transactionManager"
      class="org.springframework.jdbc.datasource.DataSourceTransactionManager">
        <property name="dataSource" ref="dataSource"/>
    </bean>
    <tx:advice id="txAdvice" transaction-manager="transactionManager">
```

```
        <!--用于配置详细的事务定义-->
        <tx:attributes>
            <!--所有以 get 开头的方法是只读的-->
            <tx:method name="get*" read-only="true" timeout="8"/>
            <!--其他方法使用默认的事务设置，指定超时时长为 5 秒-->
            <tx:method name="*" isolation="DEFAULT" propagation="REQUIRED" timeout="5"/>
        </tx:attributes>
    </tx:advice>
    <!--AOP 配置的元素-->
    <aop:config>
        <!--配置一个切入点，匹配 com.bailiban.dao.impl 包下所有以 Impl 结尾的类中所有方法的
         执行-->
        <aop:pointcut id="myPointcut" expression="execution(* com.bailiban.dao.impl.*.*(..))"/>
        <!--指定在 myPointcut 切入点应用 txAdvice 事务增强处理-->
        <aop:advisor advice-ref="txAdvice" pointcut-ref="myPointcut"/>
    </aop:config>
    <!--根据注解生成事务代理-->
    <!-- <tx:annotation-driven transaction-manager="transactionManager"/>-->
    <bean id="jdbcTemplate" class="org.springframework.jdbc.core.JdbcTemplate">
        <property name="dataSource" ref="dataSource"/>
    </bean>
</beans>
```

在上面配置文件中，我们将之前的注解驱动部分注释掉了，改为使用<tx:advisor.../>配置事务通知，并对指定的方法进行了详细的事务定义。再在<aop:config.../>中配置了一个切入点，使用<aop:advisor.../>把这个切入点与 txAdvice 绑定在一起，表示当 myPointcut 执行时，txAdvice 定义的增强处理将被织入。

### 9.2.3 事务定义器

从注解@Transactional 或者 XML 中我们都对事务进行了配置定义，如隔离级别、传播行为、事务是否只读等，这些都属于事务定义器 TransactionDefinition 的内容，其源代码如下所示。

```
package org.springframework.transaction;

import org.springframework.lang.Nullable;

public interface TransactionDefinition {
    int PROPAGATION_REQUIRED = 0;
    int PROPAGATION_SUPPORTS = 1;
    int PROPAGATION_MANDATORY = 2;
    int PROPAGATION_REQUIRES_NEW = 3;
    int PROPAGATION_NOT_SUPPORTED = 4;
```

```
    int PROPAGATION_NEVER = 5;
    int PROPAGATION_NESTED = 6;
    int ISOLATION_DEFAULT = -1;
    int ISOLATION_READ_UNCOMMITTED = 1;
    int ISOLATION_READ_COMMITTED = 2;
    int ISOLATION_REPEATABLE_READ = 4;
    int ISOLATION_SERIALIZABLE = 8;
    int TIMEOUT_DEFAULT = -1;

    default int getPropagationBehavior() {
        return 0;
    }

    default int getIsolationLevel() {
        return -1;
    }

    default int getTimeout() {
        return -1;
    }

    default boolean isReadOnly() {
        return false;
    }

    @Nullable
    default String getName() {
        return null;
    }

    static TransactionDefinition withDefaults() {
        return StaticTransactionDefinition.INSTANCE;
    }
}
```

以上就是关于事务定义器的内容，除了异常的定义，其他关于事务的定义都可以在这里完成。关于事务定义器中的隔离级别和传播行为我们会在后面单独讲解。

## 9.3 数据库事务的 ACID 特性

数据库事务正确执行的 4 个基础要素是原子性(atomicity)、一致性(consistency)、隔离性(isolation)和持久性(durability)，即 ACID。

- 原子性：一个事务内所有操作共同组成一个原子包，要么全部成功，要么全部失败。这是最基本的特性，保证了因为一些其他因素导致数据库异常，或者宕机。
- 一致性：指事务必须使数据库从一个一致性状态变换到另一个一致性状态，也就是说一个事务执行之前和执行之后都必须处于一致性状态。也就是事务的执行结果是量子化状态，而不是线性状态。
- 隔离性：指多个并发事务之间要相互隔离，互不干扰。
- 持久性：指一个事务一旦被提交，那么对数据库中的数据的改变就是永久性的，即使是在数据库系统遇到故障的情况下也不会丢失提交事务的操作。

这里的原子性、一致性和持久性都比较好理解，但是隔离性就不一样了，它涉及多个事务并发的状态。

## 9.4 事务隔离级别和传播行为

为了更好地了解事务，本节具体介绍事务的隔离级别和传播行为。

### 9.4.1 隔离级别

关于隔离级别，实际是按照 SQL 的标准规范，把隔离级别定义为 4 层，分别是：读未提交(read uncommitted)、读已提交(read committed)、可重复读(repeatable read)和串行化(serializable)。其释义如下。

- 读未提交：顾名思义，是指在一个事务处理过程里读取到了另一个未提交的事务中的数据。例如：用户 A 向用户 B 转账 100 元，但是还未提交事务，B 却已经查询到了 A 转来的 100 元钱。
- 读已提交：顾名思义，就是只能读已经提交的内容。这也是 Oracle 的默认隔离级别。可以避免“脏读”的发生，但是无法避免“不可重复读”和“幻读”。例如：A 向 B 转账 100 元钱，但是未提交；此时 B 正在查询自己的账户余额，未收到转账，同一事务下，B 再查询一次，此时 A 提交了事务，B 再次查询的结果是收到了转账的 100 元钱。虽然看似合理，但同一事务下的两次查询结果却不一样，这违背了事务的一致性。

- 可重复读：顾名思义，就是专门针对“不可重复读”这种情况而制定的隔离级别，因此它可以避免“不可重复读”这种情况。同时它也是 MySql 的默认隔离级别。但是它无法避免“幻读”的发生。因为“不可重复读”是因为两次读取之间进行了数据的修改，因此可重复读能够有效避免“不可重复读”，但却避免不了“幻读”，因为幻读是由于插入或删除数据而产生的。
- 串行化：这是数据库最高的隔离级别，这种级别下，“脏读”“不可重复读”“幻读”都可以被避免。但是执行效率差，性能开销大。

在实际工作中，注解@Transactional 隔离级别的默认值为 Isolation.DEFAULT，也就是默认的随着数据库默认值的变化而变化。

## 9.4.2 传播行为

传播行为是指方法之间的调用事务策略的问题，Spring 支持的事务传播行为如下。

- Propagation_MANDATORY：要求调用该方法的线程必须处于事务环境中，否则抛出异常。
- Propagation_NESTED：即使执行该方法的线程已处于事务环境中，也依然启动新的事务，方法在嵌套的事务里执行；即使执行该方法的线程并未处于事务环境中，也启动新的事务，然后执行该方法，此时与 Propagation_REQUIRED 相同。
- Propagation_NEVER：不允许调用该方法的线程处于事务环境中，如果调用该方法的线程处于事务中，则抛出异常。
- Propagation_NOT_SUPPORTED：如果调用该方法的线程处于事务环境中，则暂停当前事务，然后执行该方法。
- Propagation_REQUIRED：要求在事务环境中执行该方法，如果当前执行线程已处于事务环境中，则直接调用；如果当前执行线程不处于事务环境中，则启动新的事务后执行该方法。
- Propagation_REQUIRES_NEW：该方法要求在新的事务环境中执行，如果当前执行线程已处于事务环境中，则先暂停当前事务，启动新事务后执行该方法；如果当前调用线程不处于事务环境中，则启动新的事务后执行方法。
- Propagation_SUPPORTS：如果当前执行线程处于事务环境中，则使用当前事务，否则不使用事务。

上面七种传播行为中，最常用的是 Propagation_REQUIRED，也是 Spring 默认的传播行为。它比较简单，即当前如果不存在事务，就启用事务；如果存在，就沿用下来。

# 9.5 在 Spring+MyBatis 组合中使用事务

目前 Spring+MyBatis 组合应用比较流行，下面给出详细的实例。

首先，创建一个 Maven 工程，命名为 spring-mybatis-tx，然后搭建环境。代码如下所示。

## 1. spring-mybatis.xml

```
<?xml version="1.0" encoding="UTF-8"?>
<beans xmlns="http://www.springframcwork.org/schema/beans"
        xmlns:xsi="http://www.w3.org/2001/XMLSchema-instance"
        xmlns:context="http://www.springframework.org/schema/context"
        xmlns:tx="http://www.springframework.org/schema/tx"
        xsi:schemaLocation="http://www.springframework.org/schema/beans
                            http://www.springframework.org/schema/beans/spring-beans.xsd
                            http://www.springframework.org/schema/context
                            http://www.springframework.org/schema/context/spring-context.xsd
                            http://www.springframework.org/schema/tx
                            http://www.springframework.org/schema/tx/spring-tx.xsd">
    <!--读取 properties 文件-->
    <context:property-placeholder location="classpath:db.properties"/>
    <!--开启注解自动扫描-->
    <context:component-scan base-package="com.bailiban"/>
    <!--配置数据源-->
    <bean id="dataSource" class="com.alibaba.druid.pool.DruidDataSource">
        <property name="driverClassName" value="${jdbc.driver}"/>
        <property name="url" value="${jdbc.url}"/>
        <property name="username" value="${jdbc.username}"/>
        <property name="password" value="${jdbc.password}"/>
        <!--配置初始化大小、最小、最大-->
        <property name="initialSize" value="2"/>
        <property name="minIdle" value="1"/>
        <property name="maxActive" value="10"/>
        <!--配置获取连接等待超时的时间-->
        <property name="maxWait" value="60000"/>
    </bean>
    <!--配置 MyBatis -->
    <bean id="sqlSessionFactory" class="org.mybatis.spring.SqlSessionFactoryBean">
        <property name="dataSource" ref="dataSource"/>
        <property name="configLocation" value="classpath:mybatis-config.xml"/>
    </bean>
    <bean class="org.mybatis.spring.mapper.MapperScannerConfigurer">
        <property name="basePackage" value="com.bailiban.mapper"/>
        <property name="sqlSessionFactoryBeanName" value="sqlSessionFactory"/>
```

```
        </bean>
        <!--事务管理器配置数据源事务-->
        <bean id="transactionManager"
          class="org.springframework.jdbc.datasource.DataSourceTransactionManager">
              <property name="dataSource" ref="dataSource"/>
        </bean>
        <!--使用注解定义事务-->
        <tx:annotation-driven transaction-manager="transactionManager"/>
</beans>
```

## 2. mybatis-config.xml

```
<?xml version="1.0" encoding="UTF-8"?>
<!DOCTYPE configuration PUBLIC "-//mybatis.org//DTD Config 3.0//EN"
  "http://mybatis.org/dtd/mybatis-3-config.dtd">
<configuration><!--配置-->
     <settings>
          <!--指定 MyBatis 使用 log4j -->
          <setting name="logImpl" value="LOG4J"/>
          <!--开启延迟加载功能-->
          <setting name="lazyLoadingEnabled" value="true"/>
          <!--不进行完整对象的加载，属性按需加载-->
          <setting name="aggressiveLazyLoading" value="false"/>
     </settings>
     <typeAliases>
          <!--类型别名-->
          <package name="com.bailiban.entity"/>
     </typeAliases>
     <mappers>
          <mapper resource="mapper/RoleMapper.xml"/>
     </mappers>
</configuration>
```

## 3. db.properties 数据库连接配置

```
jdbc.driver=com.mysql.cj.jdbc.Driver
jdbc.url=jdbc:mysql://localhost:3306/mybatis-demo3?serverTimezone=GMT%2b8&useSSL
        =false&useUnicode=true&characterEncoding=utf-8
jdbc.username=root
jdbc.password=123456
```

## 4. log4j.properties 日志文件

```
#全局配置
log4j.rootLogger=DEBUG,stdout
#日志配置
log4j.logger.org.mybatis=DEBUG
#控制台输出配置
log4j.appender.stdout=org.apache.log4j.ConsoleAppender
```

```
log4j.appender.stdout.layout=org.apache.log4j.PatternLayout
log4j.appender.stdout.layout.ConversionPattern=%d [%t] %-5p [%c] - %m%n
```

5. Role 实体类

```
package com.bailiban.entity;

import lombok.Data;

import java.io.Serializable;

/**
 * (Role)实体类
 *
 * @author zhulang
 * @since 2020-08-06 10:38:51
 */
@Data
public class Role implements Serializable {
    private static final long serialVersionUID = -84745312631645835L;
    /**
    * 编号
    */
    private Integer id;
    /**
    * 角色名称
    */
    private String roleName;
    /**
    * 角色描述
    */
    private String roleDesp;
    /**
    * 0：不可用，1：可用
    */
    private Integer enable;

}
```

6. RoleMapper 数据库访问层

```
package com.bailiban.mapper;

import com.bailiban.entity.Role;
import org.apache.ibatis.annotations.Param;

import java.util.List;

/**
```

```
 * (Role)表数据库访问层
 *
 * @author zhulang
 * @since 2020-08-06 10:38:52
 */
public interface RoleMapper {

    /**
     * 新增数据
     *
     * @param role 实例对象
     * @return 影响行数
     */
    int insert(Role role);

}
```

### 7. RoleMapper.xml 映射文件

```
<?xml version="1.0" encoding="UTF-8"?>
<!DOCTYPE mapper PUBLIC "-//mybatis.org//DTD Mapper 3.0//EN"
 "http://mybatis.org/dtd/mybatis-3-mapper.dtd">
<mapper namespace="com.bailiban.mapper.RoleMapper">

    <!--新增所有列-->
    <insert id="insert" keyProperty="id" useGeneratedKeys="true">
        insert into tab_role(role_name, role_desp, enable)
        values (#{roleName}, #{roleDesp}, #{enable})
    </insert>
</mapper>
```

### 8. RoleService 业务层

```
package com.bailiban.service;

import com.bailiban.entity.Role;

/**
 * @author zhulang
 * @Classname RoleService
 * @Description role 业务层
 * @Date 2020/8/5 16:13
 */
public interface RoleService {

    /**
     * 插入 role
     *
     * @param role
```

```
     * @return
     */
    int insertRole(Role role);

    /**
     * 插入 role，抛出异常
     *
     * @param role
     * @return
     */
    int insertRoleWithException(Role role);

}
```

## 9. RoleServiceImpl 业务层实现类

```
package com.bailiban.service.impl;

import com.bailiban.entity.Role;
import com.bailiban.mapper.RoleMapper;
import com.bailiban.service.RoleService;
import org.springframework.beans.factory.annotation.Autowired;
import org.springframework.stereotype.Service;
import org.springframework.transaction.annotation.Isolation;
import org.springframework.transaction.annotation.Propagation;
import org.springframework.transaction.annotation.Transactional;

/**
 * @author zhulang
 * @Classname RoleServiceImpl
 * @Description role 业务层
 * @Date 2020/8/5 16:14
 */
@Service
public class RoleServiceImpl implements RoleService {
    @Autowired
    private RoleMapper roleMapper;


    @Override
    @Transactional(propagation = Propagation.REQUIRES_NEW, isolation =
                Isolation.READ_COMMITTED, rollbackFor = Exception.class)
    public int insertRole(Role role) {

        return roleMapper.insert(role);
    }

    @Override
```

```
    @Transactional(propagation = Propagation.REQUIRES_NEW, isolation =
                    Isolation.READ_COMMITTED, rollbackFor = Exception.class)
    public int insertRoleWithException(Role role) {
        roleMapper.insert(role);
        throw new RuntimeException();
    }

}
```

在 RoleServiceImpl 业务实现类中，有两个插入 Role 的方法，并且都标注了@Transactional 注解，设置了传播行为和隔离级别以及异常回滚机制。

为了方便测试，其中一个插入 Role 方法里面会抛出 RuntimeException 异常，我们设置的传播行为是 Propagation.REQUIRES_NEW。所以每当调度插入 Role 的方法时，就会产生一个新的事务，这里也可以换成其他的传播行为进行测试。

### 10. TxService 类

```
package com.bailiban.service;

import com.bailiban.entity.Role;
import org.springframework.beans.factory.annotation.Autowired;
import org.springframework.stereotype.Service;
import org.springframework.transaction.annotation.Isolation;
import org.springframework.transaction.annotation.Propagation;
import org.springframework.transaction.annotation.Transactional;

/**
 * @author zhulang
 * @Classname TxService
 * @Description 事务业务
 * @Date 2020/8/13 9:51
 */
@Service
public class TxService {
    @Autowired
    private RoleService roleService;

    @Transactional(propagation = Propagation.REQUIRED, isolation = Isolation.READ_COMMITTED,
                    rollbackFor = Exception.class)
    public void springTx1() {
        Role role = new Role();
        role.setRoleName("角色 1");
        roleService.insertRole(role);
        Role role2 = new Role();
        role2.setRoleName("角色 2");
        roleService.insertRoleWithException(role2);
```

```
    }

    @Transactional(propagation = Propagation.REQUIRED, isolation = Isolation.READ_COMMITTED,
                   rollbackFor = Exception.class)
    public void springTx2() {
        Role role = new Role();
        role.setRoleName("角色 3");
        roleService.insertRole(role);
        Role role2 = new Role();
        role2.setRoleName("角色 4");
        roleService.insertRole(role2);
        throw new RuntimeException();
    }

}
```

为了便于演示嵌套事务，我们新建了一个 TxService 类，类中两个方法也都配置了事务，因此当调用 TxService 中方法来完成插入操作时，其又会调用 RoleService 进行插入操作。它们会分别在各自的事务中进行。

上面代码中，SpringTx1 方法是插入“角色 2”时内部方法抛出异常(即插入方法中抛出异常)，SpringTx2 方法是插入“角色 4”后外部方法抛出异常(即 SpringTx2 方法中抛出异常)。

为了更好地理解传播行为，这里分析关于 Propagation.REQUIRES_NEW 和 Propagation.NESTED 的日志，它们是除了 Propagation.REQUIRED 之外，使用最多的两个传播行为。

下面在测试类中进行测试。测试代码如下所示。

```
package com.bailiban;

import com.bailiban.service.TxService;
import org.junit.Test;
import org.junit.runner.RunWith;
import org.springframework.beans.factory.annotation.Autowired;
import org.springframework.test.context.ContextConfiguration;
import org.springframework.test.context.junit4.SpringJUnit4ClassRunner;

@RunWith(SpringJUnit4ClassRunner.class)
@ContextConfiguration(locations = {"classpath:spring-mybatis.xml"})
public class AppTest {
    @Autowired
    private TxService txService;

    @Test
    public void test1() {
```

```
        txService.springTx1();
    }

    @Test
    public void test2() {
        txService.springTx2();
    }
}
```

运行测试方法 test1 和 test2，其日志截图分别如图 9-1 和图 9-2 所示。

```
Tests failed: 1 of 1 test – 1 s 531 ms
2020-08-13 15:07:42,341 [main] INFO  [com.alibaba.druid.pool.DruidDataSource] - {dataSource-1} inited
2020-08-13 15:07:42,421 [main] DEBUG [org.mybatis.spring.SqlSessionUtils] - Creating a new SqlSession❶
2020-08-13 15:07:42,435 [main] DEBUG [org.mybatis.spring.SqlSessionUtils] - Registering transaction synchronization for
 SqlSession [org.apache.ibatis.session.defaults.DefaultSqlSession@3668d4]
2020-08-13 15:07:42,451 [main] DEBUG [org.mybatis.spring.transaction.SpringManagedTransaction] - JDBC Connection
 [com.mysql.cj.jdbc.ConnectionImpl@655ef322] will be managed by Spring
2020-08-13 15:07:42,468 [main] DEBUG [com.bailiban.mapper.RoleMapper.insert] - ==>  Preparing: insert into tab_role
 (role_name, role_desp, enable) values (?, ?, ?)
2020-08-13 15:07:42,556 [main] DEBUG [com.bailiban.mapper.RoleMapper.insert] - ==> Parameters: 角色1(String), null, null
2020-08-13 15:07:42,563 [main] DEBUG [com.bailiban.mapper.RoleMapper.insert] - <==    Updates: 1
2020-08-13 15:07:42,621 [main] DEBUG [org.mybatis.spring.SqlSessionUtils] - Releasing transactional SqlSession [org
 .apache.ibatis.session.defaults.DefaultSqlSession@3668d4]
2020-08-13 15:07:42,626 [main] DEBUG [org.mybatis.spring.SqlSessionUtils] - Transaction synchronization committing❷
 SqlSession [org.apache.ibatis.session.defaults.DefaultSqlSession@3668d4]
2020-08-13 15:07:42,627 [main] DEBUG [org.mybatis.spring.SqlSessionUtils] - Transaction synchronization deregistering
 SqlSession [org.apache.ibatis.session.defaults.DefaultSqlSession@3668d4]
2020-08-13 15:07:42,628 [main] DEBUG [org.mybatis.spring.SqlSessionUtils] - Transaction synchronization closing
 SqlSession [org.apache.ibatis.session.defaults.DefaultSqlSession@3668d4]
2020-08-13 15:07:42,642 [main] DEBUG [org.mybatis.spring.SqlSessionUtils] - Creating a new SqlSession❸
2020-08-13 15:07:42,643 [main] DEBUG [org.mybatis.spring.SqlSessionUtils] - Registering transaction synchronization for
 SqlSession [org.apache.ibatis.session.defaults.DefaultSqlSession@4c4748bf]
2020-08-13 15:07:42,643 [main] DEBUG [org.mybatis.spring.transaction.SpringManagedTransaction] - JDBC Connection
 [com.mysql.cj.jdbc.ConnectionImpl@655ef322] will be managed by Spring
2020-08-13 15:07:42,644 [main] DEBUG [com.bailiban.mapper.RoleMapper.insert] - ==>  Preparing: insert into tab_role
 (role_name, role_desp, enable) values (?, ?, ?)
2020-08-13 15:07:42,645 [main] DEBUG [com.bailiban.mapper.RoleMapper.insert] - ==> Parameters: 角色2(String), null, null
2020-08-13 15:07:42,648 [main] DEBUG [com.bailiban.mapper.RoleMapper.insert] - <==    Updates: 1
2020-08-13 15:07:42,650 [main] DEBUG [org.mybatis.spring.SqlSessionUtils] - Releasing transactional SqlSession [org
 .apache.ibatis.session.defaults.DefaultSqlSession@4c4748bf]
2020-08-13 15:07:42,653 [main] DEBUG [org.mybatis.spring.SqlSessionUtils] - Transaction synchronization deregistering
 SqlSession [org.apache.ibatis.session.defaults.DefaultSqlSession@4c4748bf]
2020-08-13 15:07:42,654 [main] DEBUG [org.mybatis.spring.SqlSessionUtils] - Transaction synchronization closing
 SqlSession [org.apache.ibatis.session.defaults.DefaultSqlSession@4c4748bf]

java.lang.RuntimeException
    at com.bailiban.service.impl.RoleServiceImpl.insertRoleWithException(RoleServiceImpl.java:35) <4 internal calls>
    at org.springframework.aop.support.AopUtils.invokeJoinpointUsingReflection(AopUtils.java:344)
    at org.springframework.aop.framework.ReflectiveMethodInvocation.invokeJoinpoint(ReflectiveMethodInvocation.java:198)
    at org.springframework.aop.framework.ReflectiveMethodInvocation.proceed(ReflectiveMethodInvocation.java:163)
    at org.springframework.transaction.interceptor.TransactionAspectSupport.invokeWithinTransaction
```

图 9-1

```
Tests failed: 1 of 1 test – 1 s 556 ms
2020-08-13 15:09:31,561 [main] INFO  [com.alibaba.druid.pool.DruidDataSource] - {dataSource-1} inited
2020-08-13 15:09:31,657 [main] DEBUG [org.mybatis.spring.SqlSessionUtils] - Creating a new SqlSession❶
2020-08-13 15:09:31,670 [main] DEBUG [org.mybatis.spring.SqlSessionUtils] - Registering transaction synchronization for
 SqlSession [org.apache.ibatis.session.defaults.DefaultSqlSession@3668d4]
2020-08-13 15:09:31,686 [main] DEBUG [org.mybatis.spring.transaction.SpringManagedTransaction] - JDBC Connection
 [com.mysql.cj.jdbc.ConnectionImpl@655ef322] will be managed by Spring
2020-08-13 15:09:31,706 [main] DEBUG [com.bailiban.mapper.RoleMapper.insert] - ==>  Preparing: insert into tab_role
 (role_name, role_desp, enable) values (?, ?, ?)
2020-08-13 15:09:31,824 [main] DEBUG [com.bailiban.mapper.RoleMapper.insert] - ==> Parameters: 角色3(String), null, null
2020-08-13 15:09:31,831 [main] DEBUG [com.bailiban.mapper.RoleMapper.insert] - <==    Updates: 1
2020-08-13 15:09:31,870 [main] DEBUG [org.mybatis.spring.SqlSessionUtils] - Releasing transactional SqlSession [org
 .apache.ibatis.session.defaults.DefaultSqlSession@3668d4]
2020-08-13 15:09:31,871 [main] DEBUG [org.mybatis.spring.SqlSessionUtils] - Transaction synchronization committing❷
 SqlSession [org.apache.ibatis.session.defaults.DefaultSqlSession@3668d4]
2020-08-13 15:09:31,872 [main] DEBUG [org.mybatis.spring.SqlSessionUtils] - Transaction synchronization deregistering
 SqlSession [org.apache.ibatis.session.defaults.DefaultSqlSession@3668d4]
2020-08-13 15:09:31,873 [main] DEBUG [org.mybatis.spring.SqlSessionUtils] - Transaction synchronization closing
 SqlSession [org.apache.ibatis.session.defaults.DefaultSqlSession@3668d4]
2020-08-13 15:09:31,898 [main] DEBUG [org.mybatis.spring.SqlSessionUtils] - Creating a new SqlSession❸
2020-08-13 15:09:31,899 [main] DEBUG [org.mybatis.spring.SqlSessionUtils] - Registering transaction synchronization for
 SqlSession [org.apache.ibatis.session.defaults.DefaultSqlSession@4c4748bf]
2020-08-13 15:09:31,899 [main] DEBUG [org.mybatis.spring.transaction.SpringManagedTransaction] - JDBC Connection
 [com.mysql.cj.jdbc.ConnectionImpl@655ef322] will be managed by Spring
2020-08-13 15:09:31,899 [main] DEBUG [com.bailiban.mapper.RoleMapper.insert] - ==>  Preparing: insert into tab_role
 (role_name, role_desp, enable) values (?, ?, ?)
2020-08-13 15:09:31,900 [main] DEBUG [com.bailiban.mapper.RoleMapper.insert] - ==> Parameters: 角色4(String), null, null
2020-08-13 15:09:31,902 [main] DEBUG [com.bailiban.mapper.RoleMapper.insert] - <==    Updates: 1
2020-08-13 15:09:31,903 [main] DEBUG [org.mybatis.spring.SqlSessionUtils] - Releasing transactional SqlSession [org
 .apache.ibatis.session.defaults.DefaultSqlSession@4c4748bf]
2020-08-13 15:09:31,903 [main] DEBUG [org.mybatis.spring.SqlSessionUtils] - Transaction synchronization committing❹
 SqlSession [org.apache.ibatis.session.defaults.DefaultSqlSession@4c4748bf]
2020-08-13 15:09:31,903 [main] DEBUG [org.mybatis.spring.SqlSessionUtils] - Transaction synchronization deregistering
 SqlSession [org.apache.ibatis.session.defaults.DefaultSqlSession@4c4748bf]
2020-08-13 15:09:31,903 [main] DEBUG [org.mybatis.spring.SqlSessionUtils] - Transaction synchronization closing
 SqlSession [org.apache.ibatis.session.defaults.DefaultSqlSession@4c4748bf]

java.lang.RuntimeException
    at com.bailiban.service.TxService.springTx2(TxService.java:39)
    at com.bailiban.service.TxService$$FastClassBySpringCGLIB$$e9580b3f.invoke(<generated>)
    at org.springframework.cglib.proxy.MethodProxy.invoke(MethodProxy.java:218)
    at org.springframework.aop.framework.CglibAopProxy$CglibMethodInvocation.invokeJoinpoint(CglibAopProxy.java:771)
    at org.springframework.aop.framework.ReflectiveMethodInvocation.proceed(ReflectiveMethodInvocation.java:163)
    at org.springframework.aop.framework.CglibAopProxy$CglibMethodInvocation.proceed(CglibAopProxy.java:749)
    at org.springframework.transaction.interceptor.TransactionAspectSupport.invokeWithinTransaction
```

图 9-2

从上面两个图中可以看出，插入成功的有“角色 1”“角色 3”和“角色 4”这三个角色。接下来对此作简要分析。

在测试调用 SpringTx1 方法中，首先会开启事务，然后调用插入角色方法，因为插入“角色 1”和“角色 2”所在方法的事务传播行为是 Propagation.REQUIRES_NEW，所以它们分别在独立的新建事务中。插入“角色 2”时抛出异常，因此插入“角色 2”的方法事务被回滚，虽然抛出的异常会被外部事务感知，但是“角色 1”所处的事务是独立的。因此并不影响“角色 1”的成功插入。

同理，在测试调用 SpringTx2 中，首先会开启事务，然后调用插入角色方法，插入“角色 3”和“角色 4”会分别处在独立的新建事务中。虽然外部方法抛出异常，但是并不影响“角色 3”和“角色 4”的成功插入。

由上也可以得出结论：在外部方法开启事务的情况下，Propagation.REQUIRES_NEW 修饰的内部方法依然会单独开启独立事务，且与外围方法事务相独立，互不干扰。

下面将 RoleServiceImpl 中两个插入方法的传播行为设置为 Propagation.NESTED 进行测试。再次得到的运行日志截图分别如图 9-3 和图 9-4 所示。

```
Tests failed: 1 of 1 test – 1 s 689 ms
2020-08-13 15:25:39,152 [main] INFO  [com.alibaba.druid.pool.DruidDataSource] - {dataSource-1} inited
2020-08-13 15:25:39,316 [main] DEBUG [org.mybatis.spring.SqlSessionUtils] - Creating a new SqlSession❶
2020-08-13 15:25:39,334 [main] DEBUG [org.mybatis.spring.SqlSessionUtils] - Registering transaction synchronization for
 SqlSession [org.apache.ibatis.session.defaults.DefaultSqlSession@c055c54]
2020-08-13 15:25:39,355 [main] DEBUG [org.mybatis.spring.transaction.SpringManagedTransaction] - JDBC Connection
 [com.mysql.cj.jdbc.ConnectionImpl@544820b7] will be managed by Spring
2020-08-13 15:25:39,371 [main] DEBUG [com.bailiban.mapper.RoleMapper.insert] - ==>  Preparing: insert into tab_role
 (role_name, role_desp, enable) values (?, ?, ?)
2020-08-13 15:25:39,454 [main] DEBUG [com.bailiban.mapper.RoleMapper.insert] - ==> Parameters: 角色1(String), null, null
2020-08-13 15:25:39,459 [main] DEBUG [com.bailiban.mapper.RoleMapper.insert] - <==    Updates: 1
2020-08-13 15:25:39,487 [main] DEBUG [org.mybatis.spring.SqlSessionUtils] - Releasing transactional SqlSession [org
 .apache.ibatis.session.defaults.DefaultSqlSession@c055c54]
                                                                                                  ❷
2020-08-13 15:25:39,490 [main] DEBUG [org.mybatis.spring.SqlSessionUtils] - Fetched SqlSession [org.apache.ibatis.session
 .defaults.DefaultSqlSession@c055c54] from current transaction
2020-08-13 15:25:39,490 [main] DEBUG [com.bailiban.mapper.RoleMapper.insert] - ==>  Preparing: insert into tab_role
 (role_name, role_desp, enable) values (?, ?, ?)
2020-08-13 15:25:39,491 [main] DEBUG [com.bailiban.mapper.RoleMapper.insert] - ==> Parameters: 角色2(String), null, null
2020-08-13 15:25:39,492 [main] DEBUG [com.bailiban.mapper.RoleMapper.insert] - <==    Updates: 1
2020-08-13 15:25:39,493 [main] DEBUG [org.mybatis.spring.SqlSessionUtils] - Releasing transactional SqlSession [org
 .apache.ibatis.session.defaults.DefaultSqlSession@c055c54]
2020-08-13 15:25:39,501 [main] DEBUG [org.mybatis.spring.SqlSessionUtils] - Transaction synchronization deregistering
 SqlSession [org.apache.ibatis.session.defaults.DefaultSqlSession@c055c54]
2020-08-13 15:25:39,502 [main] DEBUG [org.mybatis.spring.SqlSessionUtils] - Transaction synchronization closing
 SqlSession [org.apache.ibatis.session.defaults.DefaultSqlSession@c055c54]

java.lang.RuntimeException
    at com.bailiban.service.impl.RoleServiceImpl.insertRoleWithException(RoleServiceImpl.java:35) <4 internal calls>
    at org.springframework.aop.support.AopUtils.invokeJoinpointUsingReflection(AopUtils.java:344)
    at org.springframework.aop.framework.ReflectiveMethodInvocation.invokeJoinpoint(ReflectiveMethodInvocation.java:198)
    at org.springframework.aop.framework.ReflectiveMethodInvocation.proceed(ReflectiveMethodInvocation.java:163)
    at org.springframework.transaction.interceptor.TransactionAspectSupport.invokeWithinTransaction
```

图 9-3

```
Tests failed: 1 of 1 test – 1 s 487 ms
2020-08-13 15:28:01,046 [main] INFO  [com.alibaba.druid.pool.DruidDataSource] - {dataSource-1} inited
2020-08-13 15:28:01,149 [main] DEBUG [org.mybatis.spring.SqlSessionUtils] - Creating a new SqlSession❶
2020-08-13 15:28:01,164 [main] DEBUG [org.mybatis.spring.SqlSessionUtils] - Registering transaction synchronization for
 SqlSession [org.apache.ibatis.session.defaults.DefaultSqlSession@c055c54]
2020-08-13 15:28:01,183 [main] DEBUG [org.mybatis.spring.transaction.SpringManagedTransaction] - JDBC Connection
 [com.mysql.cj.jdbc.ConnectionImpl@544820b7] will be managed by Spring
2020-08-13 15:28:01,221 [main] DEBUG [com.bailiban.mapper.RoleMapper.insert] - ==>  Preparing: insert into tab_role
 (role_name, role_desp, enable) values (?, ?, ?)
2020-08-13 15:28:01,391 [main] DEBUG [com.bailiban.mapper.RoleMapper.insert] - ==> Parameters: 角色3(String), null, null
2020-08-13 15:28:01,399 [main] DEBUG [com.bailiban.mapper.RoleMapper.insert] - <==    Updates: 1
2020-08-13 15:28:01,447 [main] DEBUG [org.mybatis.spring.SqlSessionUtils] - Releasing transactional SqlSession [org
 .apache.ibatis.session.defaults.DefaultSqlSession@c055c54]
                                                                                                  ❷
2020-08-13 15:28:01,450 [main] DEBUG [org.mybatis.spring.SqlSessionUtils] - Fetched SqlSession [org.apache.ibatis.session
 .defaults.DefaultSqlSession@c055c54] from current transaction
2020-08-13 15:28:01,450 [main] DEBUG [com.bailiban.mapper.RoleMapper.insert] - ==>  Preparing: insert into tab_role
 (role_name, role_desp, enable) values (?, ?, ?)
2020-08-13 15:28:01,451 [main] DEBUG [com.bailiban.mapper.RoleMapper.insert] - ==> Parameters: 角色4(String), null, null
2020-08-13 15:28:01,453 [main] DEBUG [com.bailiban.mapper.RoleMapper.insert] - <==    Updates: 1
2020-08-13 15:28:01,454 [main] DEBUG [org.mybatis.spring.SqlSessionUtils] - Releasing transactional SqlSession [org
 .apache.ibatis.session.defaults.DefaultSqlSession@c055c54]
2020-08-13 15:28:01,456 [main] DEBUG [org.mybatis.spring.SqlSessionUtils] - Transaction synchronization deregistering
 SqlSession [org.apache.ibatis.session.defaults.DefaultSqlSession@c055c54]
2020-08-13 15:28:01,456 [main] DEBUG [org.mybatis.spring.SqlSessionUtils] - Transaction synchronization closing
 SqlSession [org.apache.ibatis.session.defaults.DefaultSqlSession@c055c54]

java.lang.RuntimeException
    at com.bailiban.service.TxService.springTx2(TxService.java:39)
    at com.bailiban.service.TxService$$FastClassBySpringCGLIB$$e9580b3f.invoke(<generated>)
    at org.springframework.cglib.proxy.MethodProxy.invoke(MethodProxy.java:218)
    at org.springframework.aop.framework.CglibAopProxy$CglibMethodInvocation.invokeJoinpoint(CglibAopProxy.java:771)
    at org.springframework.aop.framework.ReflectiveMethodInvocation.proceed(ReflectiveMethodInvocation.java:163)
    at org.springframework.aop.framework.CglibAopProxy$CglibMethodInvocation.proceed(CglibAopProxy.java:749)
    at org.springframework.transaction.interceptor.TransactionAspectSupport.invokeWithinTransaction
```

图 9-4

从上面两个图中可以看出，所有角色均未成功插入。我们对此作简要的分析。

在测试调用 SpringTx1 方法中，首先会开启事务，然后调用插入角色方法，因为插入“角色 1”和“角色 2”所在方法的事务传播行为是 Propagation.NESTED，所以它们分别在独立的新建事务中，并且所处事务是外部事务的子事务，也就是说插入方法会在嵌套事务中执行。插入“角色 2”时子事务抛出异常，因此插入“角色 2”的方法事务被回滚，其抛出的异常会被外部事务感知，而外部事务感知异常会导致整体事务回滚，所以“角色 1”所处的事务也会进行回滚。这样“角色 1”和“角色 2”均未能成功插入。

同理，在测试调用 SpringTx2 方法中，外围事务抛出异常，会使整体事务均进行回滚，所以“角色 3”和“角色 4”所处的事务也会回滚。因此“角色 3”和“角色 4”均未能成功插入。

由上也可以得出结论：在外围方法开启事务的情况下，Propagation.NESTED 修饰的内部方法虽然会单独开启独立事务，但属于外围事务的子事务，外围主事务回滚，子事务一定回滚。

当然除了以上结论外，如果内部子事务回滚，是否会影响外部主事务和其他子事务呢？大家可以自行进行测试。答案是内部子事务是可以单独回滚，而不影响外部主事务和其他子事务的。

## 单元小结

- Spring 的事务策略是通过 PlatformTransactionManager 接口体现的，该接口是 Spring 事务策略的核心。
- PlatformTransactionManager 代表事务管理接口，它只要求事务管理需要提供开始事务(getTransaction())、提交事务(commit())和回滚事务(rollback())三个方法，具体实现则交给其实现类来完成。
- Spring 提供两种事务管理方式：编程式事务管理和声明式事务管理。
- Spring 事务的默认隔离级别为 Isolation.DEFAULT，表示和所使用的数据库隔离级别保持一致。
- Spring 事务的默认传播行为为 Propagation_REQUIRED。

## 单元自测

1. 下面关于 Spring 事务描述，错误的是(    )。
   A. Spring 支持可插入的事务管理器，使事务划分更轻松，同时无需处理底层的问题
   B. Spring 事务管理的通用抽象层还包括 JTA 策略和一个 JDBC DataSource
   C. 与 JTA 或 EJB CMT 一样，Spring 的事务支持依赖于 Java EE 环境
   D. Spring 事务语义通过 AOP 应用于 POJO，通过 XML 或 Java SE 5 注释进行配置
2. 下面关于编程式事务管理说法，正确的是(    )。
   A. 通过编程的方式管理事务，带来极大的灵活性，且便于维护
   B. 编程式事务管理可以将业务代码和事务管理分离
   C. Spring 的编程式事务还可以通过 TransactionTemplate 类来完成
   D. 对于编程式事务管理，程序无法直接获取容器中的 transactionManager Bean
3. Spring 框架的事务管理具有的优点有(    )。
   A. 它为不同的事务 API 如 JTA、JDBC、Hibernate、JPA 和 JDO，提供一个不变的编程模式
   B. 它为编程式事务管理提供了一套简单的 API
   C. 它支持声明式事务管理
   D. 它和 Spring 各种数据访问抽象层进行了很好的集成
4. 下面关于@Transactional 注解描述，正确的是(    )。
   A. 只能作用在具体的方法上
   B. 只能作用在类上
   C. 使用注解配置事务，还需要配置注解驱动
   D. 必须指定传播行为
5. 下面关于 Spring 事务自动回滚的说法，正确的是(    )。
   A. 在默认情况下，只有当方法引发运行时异常时，才会进行自动回滚
   B. 在默认情况下，如果事务抛出 checked 异常，会导致事务回滚
   C. 可以通过 rollback-for 属性强制 Spring 遇到特定 checked 异常时自动回滚事务
   D. 可以通过 no-rollback-for 属性强制 Spring 遇到特定异常时不回滚事务

## 上机实战

### 上机目标

掌握 Spring 事务的传播行为和声明式事务注解开发。

### 上机练习

### 练习：完成本章关于 Propagation.NESTED 传播行为的测试。

**【问题描述】**

如果内部子事务回滚，是否会影响外围主事务和其他子事务？

**【问题分析】**

在本章测试中虽然也测试了子事务抛出异常的情况，得出的测试结果是“角色 1”和“角色 2”均未成功插入。那么是否表示子事务不能独自回滚？

**【参考步骤】**

(1) 在 TxService 类中再添加一个方法。对子事务抛出的异常进行捕获，代码如下。

```
@Transactional(propagation = Propagation.REQUIRED, isolation = Isolation.READ_COMMITTED,
                    rollbackFor = Exception.class)
public void springTx3() {
        Role role = new Role();
        role.setRoleName("角色 1");
        roleService.insertRole(role);
        Role role2 = new Role();
        role2.setRoleName("角色 2");
      try {
                roleService.insertRoleWithException(role2);
        }catch (Exception e){
                System.out.println("捕获子事务抛出的异常，方法回滚！");
      }
}
```

(2) 在测试类中创建 test3 对其进行测试。代码如下。

```
@Test
public void test3() {
        txService.springTx3();
}
```

(3) 测试结果显示“角色 1”插入成功，“角色 2”未插入。

(4) 由此可以证明：内部子事务是可以单独回滚而不影响外部主事务和其他子事务的。

# Spring MVC的开发流程

## 课程目标

❖ 掌握 MVC 框架的特点

❖ 掌握 Spring MVC 框架的架构设计

❖ 学会 Spring MVC 的入门使用

❖ 掌握 Spring MVC 的开发流程

### 简介

Spring Web MVC(简称 Spring MVC)是 Spring 提供给 Web 应用的框架设计，除了 Spring MVC 外，比较优秀的 Web 应用框架还有 Struts 2，但由于其配置繁琐且曾曝出过漏洞，导致 Struts 2 也被慢慢淘汰。目前主流的 Web 应用框架便是 Spring MVC。本单元将介绍它的初始化和流程。

## 10.1 MVC 设计概述

MVC(Model-View-Controller)模式是软件工程中的一种软件架构模式，把软件系统分为三个基本部分：模型(Model)、视图(View)和控制器(Controller)。MVC 可对程序的后期维护和扩展提供方便，也为程序某些部分的重用提供了方便。

MVC 设计模式并不是 Java Web 应用的专属，几乎现在所有 B/S 结构的软件都采用了 MVC 设计模式。在早期的 Java Web 开发中，主要是 JSP+Java Bean 模式，我们称之为 Model 1，如图 10-1 所示。服务器端只有 JSP 页面，所有操作都在 JSP 页面中，严重的耦合度增加了开发的难度，对后期的维护和扩展极其不利。

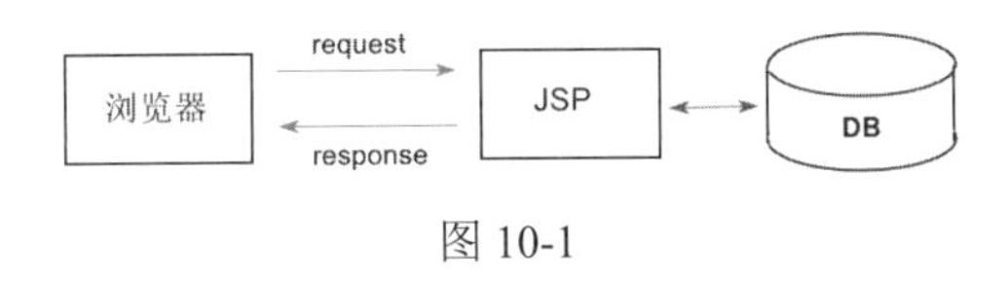


图 10-1

正因为存在这种弊端，后期在 Model 1 的基础上进行了改进，我们称之为 Model 的第二代，如图 10-2 所示。Model 2 把业务逻辑的内容放到了 Java Bean 中，而 JSP 页面负责显示以及请求调度的工作。虽然有所改进，但是 JSP 中把视图工作和请求调度的工作耦合在一起了。

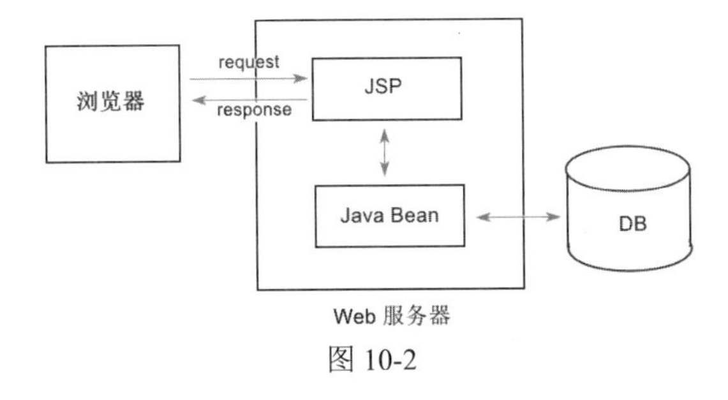


图 10-2

到了后来，Model 2 模式是相对比较完整的 MVC 设计结构，如图 10-3 所示。

- JSP：视图层，负责接收和显示数据给用户。

- Servlet：控制层，负责找到合适的模型对象来处理业务逻辑，以转发到合适的视图。
- JavaBean：模型层，完成具体的业务工作。

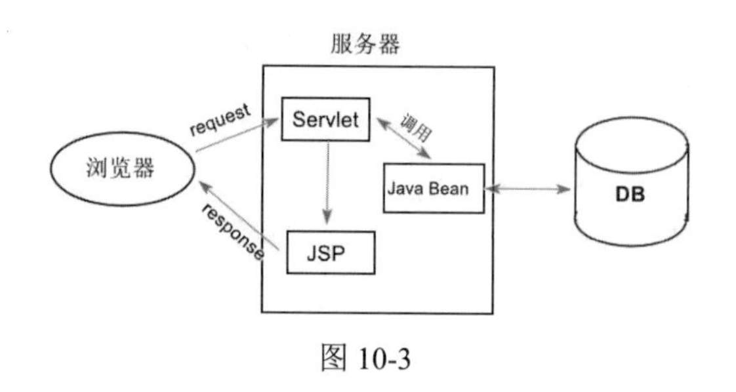


图 10-3

随着互联网的快速发展，手机端逐步兴起，页面和后台的数据交互更多的是采用 JSON 数据，前后端分离已经成为一种发展的趋势，这样后台代码可以得到重用，因此 Spring MVC 框架便是主流的选择。

Spring MVC 的核心在于其流程，这是使用 Spring MVC 框架的基础，Spring MVC 是一种基于 Servlet 的技术，它提供了核心控制器 DispatcherServlet 和相关的组件，并制定了松散的结构，以适应各种灵活的需求。其流程图如图 10-4 所示。

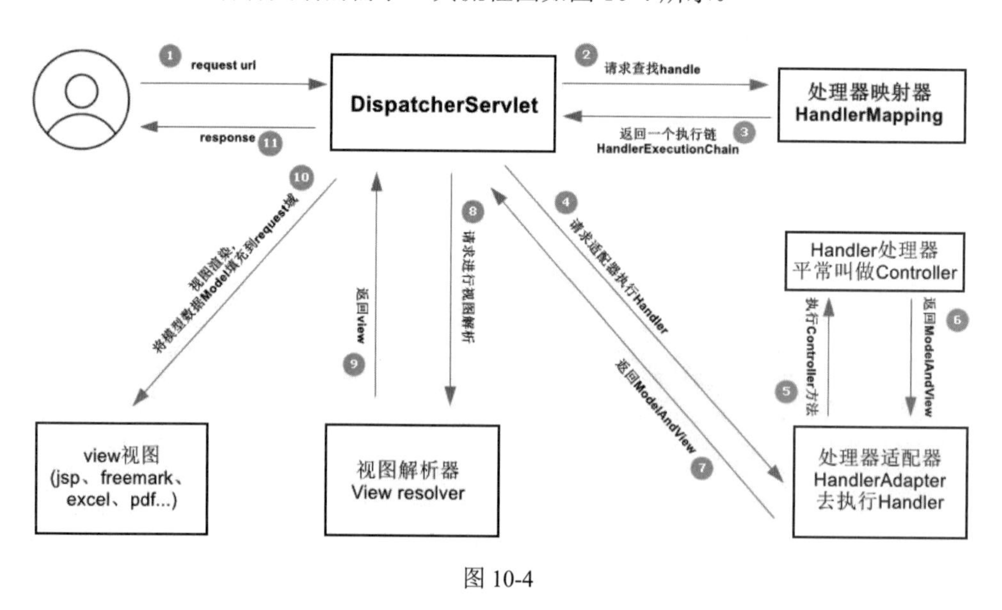


图 10-4

图 10-4 中标注了 Spring MVC 的服务流程及各个组件运行的顺序，这是 Spring MVC 的核心。

首先，Spring MVC 框架是围绕着 DispatcherServlet 工作的，这个是其核心类，其实它本质是一个 Servlet，因此它可以拦截 HTTP 发送过来的请求，在 Servlet 初始化时，Spring MVC 会根据配置，获取配置信息，从而得到统一资源标识符(URI，Uniform Resource Identifier)和处理器(Handler)之间的映射关系(HandlerMapping)。为了使用更加灵活，且能增强功能，Spring MVC 还会给处理器加入拦截器，所以还可以在处理器执行前后加入自己的代码，这样就构成了一个处理器的执行链(Handler Execution Chain)，并且根据上下文初始化视图解析器等内容，当处理器返回的时候就可以通过视图解析器定位视图，然后将数据模型渲染到视图中，以响应用户的请求。

具体的步骤如下：

第一步：发起请求到前端控制器(DispatcherServlet)。

第二步：前端控制器请求 HandlerMapping 查找 Handler (可以根据 xml 配置、注解进行查找)。

第三步：处理器映射器 HandlerMapping 向前端控制器返回 Handler，HandlerMapping 会把请求映射为 HandlerExecutionChain 对象(包含一个 Handler 处理器(页面控制器)对象，多个 HandlerInterceptor 拦截器对象)，通过这种策略模式，很容易添加新的映射策略。

第四步：前端控制器调用处理器适配器去执行 Handler。

第五步：处理器适配器 HandlerAdapter 将会根据适配的结果执行 Handler。

第六步：Handler 执行完成给适配器返回 ModelAndView。

第七步：处理器适配器向前端控制器返回 ModelAndView (ModelAndView 是 Spring MVC 框架的一个底层对象，包括 Model 和 View)。

第八步：前端控制器请求视图解析器进行视图解析 (根据逻辑视图名解析成真正的视图(.jsp))，通过这种策略很容易更换其他视图技术，只更改视图解析器即可。

第九步：视图解析器向前端控制器返回 View。

第十步：前端控制器进行视图渲染 (视图渲染将模型数据(在 ModelAndView 对象中)填充到 request 域)。

第十一步：前端控制器向用户响应结果。

以上就是一个 Spring MVC 完整的流程，它是一个松散的结构，所以可以满足各类请求的需要，为此它也实现了大部分请求所需的类库，拥有较为丰富的类库供使用，所以流程中大部分组件并不需要用户去实现。

## 10.2 Spring MVC 入门实例

作为 Spring MVC 入门，由于在实际开发中 Spring MVC 的配置主要以 XML 为主，因此这里选择使用 XML 配置的方式。

首先创建一个 Maven，工程名为 springmvc-demo，在选择 Maven 骨架的时候，选择 maven-archetype-webapp，创建成功后构建如图 10-5 所示的目录结构。

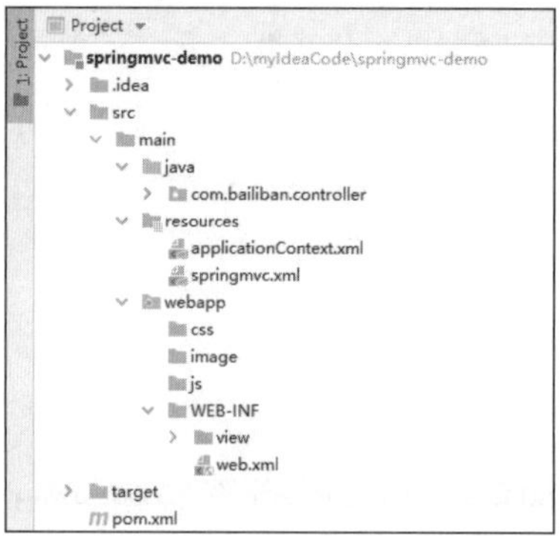

图 10-5

(1) 首先为 pom 文件添加依赖坐标，pom 文件的完整内容如下。

```
<?xml version="1.0" encoding="UTF-8"?>

<project xmlns="http://maven.apache.org/POM/4.0.0"
 xmlns:xsi="http://www.w3.org/2001/XMLSchema-instance"
         xsi:schemaLocation="http://maven.apache.org/POM/4.0.0
                        http://maven.apache.org/xsd/maven-4.0.0.xsd">
    <modelVersion>4.0.0</modelVersion>

    <groupId>com.bailiban</groupId>
    <artifactId>springmvc-demo</artifactId>
    <version>1.0-SNAPSHOT</version>
    <packaging>war</packaging>

    <name>springmvc-demo</name>

    <properties>
        <project.build.sourceEncoding>UTF-8</project.build.sourceEncoding>
        <maven.compiler.source>1.8</maven.compiler.source>
        <maven.compiler.target>1.8</maven.compiler.target>
    </properties>

    <dependencies>
        <dependency>
            <groupId>org.springframework</groupId>
            <artifactId>spring-webmvc</artifactId>
            <version>5.2.7.RELEASE</version>
        </dependency>
        <dependency>
            <groupId>org.projectlombok</groupId>
            <artifactId>lombok</artifactId>
            <version>1.18.12</version>
            <scope>provided</scope>
        </dependency>
```

```
        <dependency>
            <groupId>log4j</groupId>
            <artifactId>log4j</artifactId>
            <version>1.2.17</version>
        </dependency>
    </dependencies>
    <build>
        <plugins>
            <plugin>
                <groupId>org.apache.tomcat.maven</groupId>
                <artifactId>tomcat8-maven-plugin</artifactId>
                <version>3.0-r1756463</version>
                <configuration>
                    <server>tomcat8</server>
                    <port>8080</port>
                    <path>/</path>
                </configuration>
            </plugin>
        </plugins>
    </build>
    <pluginRepositories>
        <pluginRepository>
            <id>alfresco-public</id>
            <url>https://artifacts.alfresco.com/nexus/content/groups/public</url>
        </pluginRepository>
        <pluginRepository>
            <id>alfresco-public-snapshots</id>
            <url>https://artifacts.alfresco.com/nexus/content/groups/public-snapshots</url>
            <snapshots>
                <enabled>true</enabled>
                <updatePolicy>daily</updatePolicy>
            </snapshots>
        </pluginRepository>
        <pluginRepository>
            <id>beardedgeeks-releases</id>
            <url>http://beardedgeeks.googlecode.com/svn/repository/releases</url>
        </pluginRepository>
    </pluginRepositories>
</project>
```

pom 文件中关于 Spring 的 jar 包虽然只添加了 spring-webmvc 的依赖坐标，但由于它同时会依赖 Spring 其他的一些 jar 包，所以也会把相关联的依赖添加到项目资源库中(External Libraries)。

除了依赖坐标外，还使用了<plugins></plugins>标签，添加了 Tomcat 8 插件，同时给它配置了端口号 8080，以及部署的项目路径为/，Maven 中的<pluginRepositories>...</pluginRepositories>

是用来配置插件地址的，因为 Maven 的所有功能都是使用插件来实现功能，因此需要从特定的地址下载插件包。

(2) 配置 web.xml。

```
<?xml version="1.0" encoding="UTF-8"?>
<web-app xmlns="http://xmlns.jcp.org/xml/ns/javaee"
         xmlns:xsi="http://www.w3.org/2001/XMLSchema-instance"
         xsi:schemaLocation="http://xmlns.jcp.org/xml/ns/javaee
                             http://xmlns.jcp.org/xml/ns/javaee/web-app_4_0.xsd"
         version="4.0">
    <!--配置 Spring IoC 配置文件路径-->
    <context-param>
        <param-name>contextConfigLocation</param-name>
        <param-value>classpath*:applicationContext.xml</param-value>
    </context-param>
    <!--配置 ContextLoaderListener 用以初始化 Spring IoC 容器-->
    <listener>
        <listener-class>org.springframework.web.context.ContextLoaderListener</listener-class>
    </listener>
    <!--配置核心控制器-->
    <servlet>
        <!--名称-->
        <servlet-name>dispatcher</servlet-name>
        <!--Servlet 类-->
        <servlet-class>org.springframework.web.servlet.DispatcherServlet</servlet-class>
        <init-param>
            <!--Spring MVC  配置参数文件的位置-->
            <param-name>contextConfigLocation</param-name>
            <!--(可不配置)默认名称为 ServletName-servlet.xml，即 dispatcher-servlet.xml-->
            <param-value>classpath*:springmvc.xml</param-value>
        </init-param>
        <!--动顺序，数字越小，启动越早，非 0 正整数-->
        <load-on-startup>1</load-on-startup>
    </servlet>
    <!--Servlet 拦截配置-->
    <servlet-mapping>
        <servlet-name>dispatcher</servlet-name>
        <url-pattern>/</url-pattern>
    </servlet-mapping>
</web-app>
```

关于 web.xml 中的一些配置解释如下。

① contextConfigLocation：用于指定配置文件的位置。如果有多个配置文件，以逗号隔开；如果没有指定配置文件，Spring 监听器会自动查找/WEB-INF/路径下的 applicationContext.xml 配置文件。

② ContextLoaderListener：它是 Spring 提供的监听器，实现了 ServletContextListener 接口。由它来查找 Spring 的配置文件，完成对 Spring 的初始化。

③ DispatcherServlet：它是 Spring 的核心控制器，其初始化属性配置用于指定 springmvc 的配置文件位置。如果没有进行初始化配置，其默认的配置文件在/WEB-INF/路径下，名称为配置的 servlet-name 连上-servlet.xml。例如 servlet-name 为 dispatcher，则配置文件名称为 dispatcher-servlet.xml。

④ servlet-mapping：它是 servlet 拦截配置，servlet-name 需要和上面配置的 servlet 中的 servlet-name 一致。<url-pattern>是拦截指定形式的请求，例如这里配置的是“/”，则会拦截所有路径型的 url；如果配置的是*.do，则会拦截所有以后缀“do”结尾的请求。

(3) 在 resources 下创建 springmvc.xml，内容如下。

```
<?xml version="1.0" encoding="UTF-8"?>
<beans xmlns="http://www.springframework.org/schema/beans"
        xmlns:mvc="http://www.springframework.org/schema/mvc"
        xmlns:xsi="http://www.w3.org/2001/XMLSchema-instance"
        xmlns:aop="http://www.springframework.org/schema/aop"
        xmlns:context="http://www.springframework.org/schema/context"
        xsi:schemaLocation="
                        http://www.springframework.org/schema/beans
                        http://www.springframework.org/schema/beans/spring-beans.xsd
                        http://www.springframework.org/schema/context
                        http://www.springframework.org/schema/context/spring-context.xsd
                        http://www.springframework.org/schema/mvc
                        http://www.springframework.org/schema/mvc/spring-mvc.xsd
                        http://www.springframework.org/schema/aop
                        http://www.springframework.org/schema/aop/spring-aop.xsd">
    <!--使用注解驱动-->
    <mvc:annotation-driven/>
    <!--定义扫描装载的包-->
    <context:component-scan base-package="com.bailiban.controller"/>
    <!--配置视图解析器-->
    <bean id="view" class="org.springframework.web.servlet.view.InternalResourceViewResolver">
        <property name="prefix" value="/WEB-INF/view/"/>
        <property name="suffix" value=".jsp"/>
    </bean>
    <!--不拦截静态资源-->
    <mvc:resources location="/css/" mapping="/css/**"/>
    <mvc:resources location="/js/" mapping="/js/**"/>
    <mvc:resources location="/image/" mapping="/image/**"/>

</beans>
```

关于 springmvc.xml 中的一些配置解释如下。

① <mvc:annotation-driven/>：表示使用注解驱动 Spring MVC。

② <context:component-scan/>：定义要扫描注解的包。

③ InternalResourceViewResolver：定义视图解析器，解析器中定义了前缀和后缀，这样视图就知道去 Web 工程的/WEB-INF/view 文件夹中找到对应的 JSP 文件作为视图响应用户请求。

④ <mvc:resources/>：为了避免拦截器将一些静态资源也给拦截了，通过属性 location 和 mapping 就可以让拦截器对指定目录下的资源不进行拦截。

(4) 配置 applicationContext.xml。

```
<?xml version="1.0" encoding="UTF-8"?>
<beans xmlns="http://www.springframework.org/schema/beans"
        xmlns:xsi="http://www.w3.org/2001/XMLSchema-instance"
        xsi:schemaLocation="http://www.springframework.org/schema/beans
                            http://www.springframework.org/schema/beans/spring-beans.xsd">

</beans>
```

关于目前的入门案例，applicationContext.xml 中不用配置任何内容。

(5) 在包 com.bailiban.controller 下创建 IndexController 类，代码如下。

```
package com.bailiban.controller;

import org.springframework.stereotype.Controller;
import org.springframework.web.bind.annotation.RequestMapping;

/**
 * @author zhulang
 * @Classname IndexController
 * @Description index 控制器
 * @Date 2020/8/14 17:32
 */
@Controller
public class IndexController {
    @RequestMapping("/index")
    public String index() {
        return "index";
    }
}
```

上面的代码十分简单，@Controller 注解当前类是一个控制器。Spring MVC 扫描的时候就会把它作为控制器加载进来。@RequestMapping 指定了对应的请求 URI。当有对应的请求进来时，会返回一个字符串“index”，作为视图名称，表示不需要数据模型。

由于在 springmvc.xml 配置中配置了视图解析器，前缀为/WEB-INF/view，后缀为.jsp，所以最终的响应视图便是/WEB-INF/view/index.jsp。

(6) 在 WEB-INF 下创建 View 目录，再在 View 目录下创建 index.jsp 文件，内容如下。

```
<html>
<body>
<h2>Hello Spring MVC!</h2>
</body>
</html>
```

(7) 最后启动服务器 Tomcat。

在 Maven 工程中 Idea 提供了快捷的 Maven 执行命令，如图 10-6 所示，直接选择命令 tomcat8:run-war 来启动 Web 项目。

由于要下载一些插件资源，初次启动会比较慢。等待成功启动后，在浏览器输入 localhost:8080/index，就能看到相应的页面结果，如图 10-7 所示。

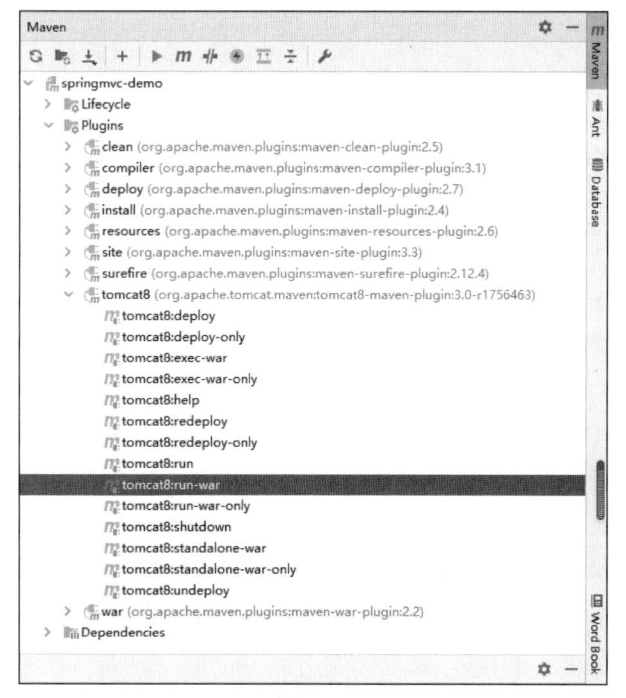


图 10-6

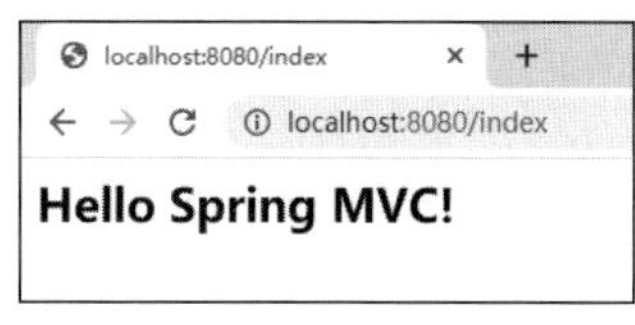


图 10-7

# 10.3 Spring MVC 具体开发流程

上一节中，我们完成了 Spring MVC 的入门实例开发，发现开发 Spring MVC 流程并不困难，但还需要掌握 Spring MVC 的组件和流程。

在目前的主流开发中，大部分都会采用注解的开发方式，除了一些配置采用 XML 编写外。例如，这里主要使用了@Controller 注解及@RequestMapping 注解。这些都是 Spring MVC 中最常用的注解。

## 10.3.1 @RequestMapping 注解

在 Spring MVC 应用程序中，RequestDispatcher 这个 Servlet 负责将进入的 HTTP 请求路由到控制器的处理方法。

在对 Spring MVC 进行配置的时候，需要指定请求与处理方法之间的映射关系。要配置 Web 请求的映射，就需要用到@RequestMapping 注解。接下来先看它的源码。

```
package org.springframework.web.bind.annotation;

import java.lang.annotation.Documented;
import java.lang.annotation.ElementType;
import java.lang.annotation.Retention;
import java.lang.annotation.RetentionPolicy;
import java.lang.annotation.Target;
import org.springframework.core.annotation.AliasFor;

@Target({ElementType.TYPE, ElementType.METHOD})
@Retention(RetentionPolicy.RUNTIME)
@Documented
@Mapping
public @interface RequestMapping {
    String name() default "";

    @AliasFor("path")
    String[] value() default {};

    @AliasFor("value")
    String[] path() default {};

    RequestMethod[] method() default {};

    String[] params() default {};
```

```
        String[] headers() default {};

        String[] consumes() default {};

        String[] produces() default {};
    }
```

从源码中可以看出，@RequestMapping 注解可以作用在控制器类上，也可以作用在控制器类中的方法上。这里最常用到的就是请求路径和请求类型，其中请求路径可以是字符串，也可以是数组。而最常用的请求类型有 Http 的 GET 请求和 POST 请求，代码如下所示。

```
package com.bailiban.controller;

import org.springframework.stereotype.Controller;
import org.springframework.web.bind.annotation.RequestMapping;
import org.springframework.web.bind.annotation.RequestMethod;

/**
 * @author zhulang
 * @Classname IndexController
 * @Description index 控制器
 * @Date 2020/8/14 17:32
 */
@Controller
@RequestMapping("/springmvc")
public class IndexController {
    @RequestMapping(value = "/index", method = RequestMethod.GET)
    public String index() {
        return "index";
    }
}
```

对 IndexController 进行修改，在控制器 IndexController 类中加上@RequestMapping 注解，同时在 index 方法上的@RequestMapping 中限定请求类型是 GET 请求。如果想要再次访问 index.jsp，就应该在浏览器中输入 localhost:8080/springmvc/index，才能得到响应。

## 10.3.2 控制器的开发

控制器开发是 Spring MVC 的核心内容，主要分为以下三个步骤。

(1) 获取请求参数。

(2) 处理业务逻辑。

(3) 绑定模型和视图。

下面分别进行讲解。

### 1. 获取请求参数

在学习 Servlet 的时候，大家可能习惯通过 request 或者 session 来获取 HTTP 请求参数。在 Spring MVC 中提供了更好的方式来获取请求参数，例如注解@RequestParam。使用@RequestParam 注解可以将请求参数绑定到控制器的方法参数上。代码如下所示。

```
@RequestMapping(value = "/doLogin", method = RequestMethod.POST)
public String login(@RequestParam("username") String name, String password) {
    System.out.println(name);
    System.out.println(password);
    return "success";
}
```

登录页面中会提交两个请求参数，分别为 username 和 password。如果没有使用@RequestParam 注解，就要求方法中的参数名称和请求参数名必须一致。当不一致时，就可以通过@RequestParam 将请求的参数绑定到方法参数上。

在默认情况下对于注解@RequestParam 的参数而言，它要求参数不能为空，否则会抛出异常。但有时候允许出现参数为空，自行赋予默认值的情况。此时可以通过如下配置来实现。代码如下所示。

```
@RequestMapping(value = "/doLogin", method = RequestMethod.POST)
public String login(@RequestParam(value = "username",required = false,defaultValue = "bailiban") String
name, String password) {
    System.out.println(name);
    System.out.println(password);
    return "success";
}
```

关于@RequestParam 中的属性配置解释如下。

- required：值类型为布尔值，默认值为 true，表示不允许参数为空，如果要允许为空，则设置为 false。
- defaultValue：当参数为空时候的默认值。例如，这里当 username 为空时，默认值为“bailiban”。

关于 Controller 参数的获取，这里只作了简单介绍，在下一单元中再详细介绍。

### 2. 业务逻辑处理和视图绑定

关于业务逻辑处理，基本上都会跟数据库有关联，这里使用 XML 配置，在 applicationContext.xml 中配置关于数据库的部分信息，使用 MyBatis 作为持久层框架。配置代码如下所示。

```
<?xml version="1.0" encoding="UTF-8"?>
<beans xmlns="http://www.springframework.org/schema/beans"
        xmlns:xsi="http://www.w3.org/2001/XMLSchema-instance"
        xmlns:context="http://www.springframework.org/schema/context"
        xmlns:tx="http://www.springframework.org/schema/tx"
        xsi:schemaLocation="http://www.springframework.org/schema/beans
                            http://www.springframework.org/schema/beans/spring-beans.xsd
                            http://www.springframework.org/schema/context
                            http://www.springframework.org/schema/context/spring-context.xsd
                            http://www.springframework.org/schema/tx
                            http://www.springframework.org/schema/tx/spring-tx.xsd">
    <!--读取 properties 文件-->
    <context:property-placeholder location="classpath:properties/db.properties"/>
    <!--开启注解自动扫描-->
    <context:component-scan base-package="com.bailiban"/>
    <!--配置数据源-->
    <bean id="dataSource" class="com.alibaba.druid.pool.DruidDataSource">
        <property name="driverClassName" value="${jdbc.driver}"/>
        <property name="url" value="${jdbc.url}"/>
        <property name="username" value="${jdbc.username}"/>
        <property name="password" value="${jdbc.password}"/>
        <!--配置初始化大小、最小、最大-->
        <property name="initialSize" value="2"/>
        <property name="minIdle" value="1"/>
        <property name="maxActive" value="10"/>
        <!--配置获取连接等待超时的时间-->
        <property name="maxWait" value="60000"/>
    </bean>
    <!--配置 MyBatis -->
    <bean id="sqlSessionFactory" class="org.mybatis.spring.SqlSessionFactoryBean">
        <property name="dataSource" ref="dataSource"/>
        <property name="configLocation" value="classpath:mybatis/mybatis-config.xml"/>
    </bean>
    <bean class="org.mybatis.spring.mapper.MapperScannerConfigurer">
        <property name="basePackage" value="com.bailiban.mapper"/>
        <property name="sqlSessionFactoryBeanName" value="sqlSessionFactory"/>
    </bean>
    <!--事务管理器配置数据源事务-->
    <bean id="transactionManager"
     class="org.springframework.jdbc.datasource.DataSourceTransactionManager">
        <property name="dataSource" ref="dataSource"/>
    </bean>
    <!--使用注解定义事务-->
    <tx:annotation-driven/>
</beans>
```

关于这些配置中所需要的接口和类，已经在上一单元中进行了介绍，这里就不再展示相关代码。接下来将演示如何查询所有角色，控制器代码如下。

```
package com.bailiban.controller;

import com.bailiban.entity.Role;
import com.bailiban.service.RoleService;
import org.springframework.beans.factory.annotation.Autowired;
import org.springframework.stereotype.Controller;
import org.springframework.ui.Model;
import org.springframework.web.bind.annotation.RequestMapping;

import java.util.List;

/**
 * @author zhulang
 * @Classname RoleController
 * @Description role 控制器
 * @Date 2020/8/18 10:22
 */
@Controller
@RequestMapping("/role")
public class RoleController {
    @Autowired
    private RoleService roleService;

    @RequestMapping("/all")
    public String findAll(Model model) {
        List<Role> roles = roleService.findAll();
        model.addAttribute("roles", roles);
        return "role-list";
    }
}
```

上面代码中给数据模型对象添加了数据，这样在 jsp 页面中就可以通过 el 表达式${}获取到。

RoleService 中的方法也比较简单，其对应的实现代码如下。

```
package com.bailiban.service.impl;

import com.bailiban.entity.Role;
import com.bailiban.mapper.RoleMapper;
import com.bailiban.service.RoleService;
import org.springframework.beans.factory.annotation.Autowired;
import org.springframework.stereotype.Service;
import org.springframework.transaction.annotation.Transactional;
```

```
import java.util.List;

/**
 * @author zhulang
 * @Classname RoleServiceImpl
 * @Description 业务实现类
 * @Date 2020/8/18 10:14
 */
@Service
@Transactional(rollbackFor = Exception.class)
public class RoleServiceImpl implements RoleService {
    @Autowired
    private RoleMapper roleMapper;

    @Override
    public List<Role> findAll() {
        return roleMapper.findAll();
    }
}
```

这样就完成了业务逻辑，但是并没有实现视图渲染，也就是还需要把查询出来的数据，通过某种方式渲染到视图中，展示给用户。

## 10.3.3　视图渲染

因为这里使用的视图是 JSP，因此可以通过 JSTL 在 JSP 中展示数据模型。

在控制器中使用的视图名是 role-list，因此最终会响应到/WEB-INF/view/role-list.jsp 文件中。其代码如下所示。

```
<%@ page contentType="text/html;charset=UTF-8" %>
<%@taglib prefix="c" uri="http://java.sun.com/jsp/jstl/core" %>
<!DOCTYPE html>
<html>
<head>
    <!-- 页面 meta -->
    <meta charset="utf-8">
    <meta http-equiv="X-UA-Compatible" content="IE=edge">
    <title>角色展示</title>
    <link href="${pageContext.request.contextPath}/css/bootstrap.min.css" rel="stylesheet">
    <script src="${pageContext.request.contextPath}/js/jquery-3.2.1.min.js"></script>
    <script src="${pageContext.request.contextPath}/js/bootstrap.min.js"></script>
</head>
<body>
<div class="container">
    <table class="table">
```

```
            <thead>
            <tr>
                <th>ID</th>
                <th>名称</th>
                <th>描述</th>
                <th>状态</th>
            </tr>
            </thead>
            <c:forEach items="${roles}" var="role">
                <tr>
                    <td>${role.id}</td>
                    <td>${role.roleName}</td>
                    <td>${role.roleDesp}</td>
                    <td>${role.enable==1?"可用":"不可用"}</td>
                </tr>
            </c:forEach>
        </table>
    </div>
    </body>
    </html>
```

在 role-list.jsp 页面中，我们使用 Bootstrap 框架中的一点内容来简单美化视图页面。通过 EL 表达式${roles}读取出 role 集合，然后通过 c 标签 forEach 进行遍历。启动服务器，在浏览器中输入 localhost:8080/role/all，即可得到相应的结果，如图 10-8 所示。

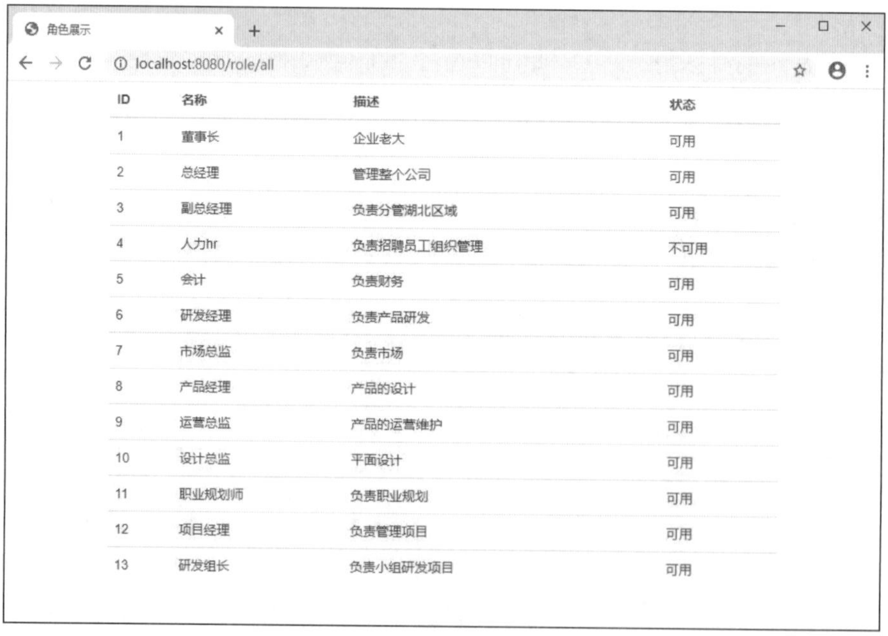

角色展示

localhost:8080/role/all

| ID | 名称 | 描述 | 状态 |
|---|---|---|---|
| 1 | 董事长 | 企业老大 | 可用 |
| 2 | 总经理 | 管理整个公司 | 可用 |
| 3 | 副总经理 | 负责分管湖北区域 | 可用 |
| 4 | 人力hr | 负责招聘员工组织管理 | 不可用 |
| 5 | 会计 | 负责财务 | 可用 |
| 6 | 研发经理 | 负责产品研发 | 可用 |
| 7 | 市场总监 | 负责市场 | 可用 |
| 8 | 产品经理 | 产品的设计 | 可用 |
| 9 | 运营总监 | 产品的运营维护 | 可用 |
| 10 | 设计总监 | 平面设计 | 可用 |
| 11 | 职业规划师 | 负责职业规划 | 可用 |
| 12 | 项目经理 | 负责管理项目 | 可用 |
| 13 | 研发组长 | 负责小组研发项目 | 可用 |

图 10-8

当然在如今实际开发中，随着前端技术的快速迭代发展，JSP 也已经不再是首选技术。

并且前后端分离的状态下，后台更多的是返回 JSON 格式的数据给前台页面。Spring MVC 也可以很好地实现。将控制器中的 findAll 方法修改为如下代码。

```
@RequestMapping("/all")
@ResponseBody
public List<Role> findAll() {
    List<Role> roles = roleService.findAll();
    return roles;
}
```

Spring MVC 提供了注解@ResponseBody，它可以将 Controller 的方法返回的对象通过适当的转换器转换为指定的格式，之后写入到 response 对象的 body 区，以返回 JSON 数据或者 XML 数据。因此还需要配置转换器，在 springmvc.xml 中加入如下代码。

```
<mvc:annotation-driven>
    <mvc:message-converters>
        <bean class="org.springframework.http.converter.json.MappingJackson2HttpMessageConverter"/>
    </mvc:message-converters>
</mvc:annotation-driven>
```

此外它依赖于 Jackson 2 包，因此需要在 pom 文件中加入如下依赖坐标。

```
<dependency>
    <groupId>com.fasterxml.jackson.core</groupId>
    <artifactId>jackson-databind</artifactId>
    <version>2.9.5</version>
</dependency>
<dependency>
    <groupId>com.fasterxml.jackson.core</groupId>
    <artifactId>jackson-annotations</artifactId>
    <version>2.9.5</version>
</dependency>
```

再次刷新浏览器，可以得到如图 10-9 所示的响应结果。

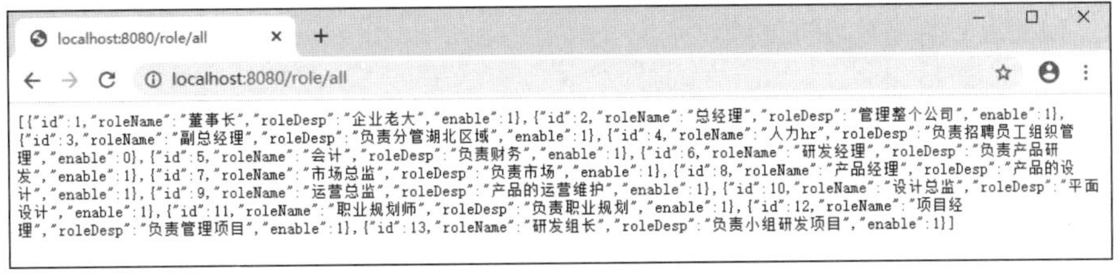


图 10-9

这样前端页面通过调用后台接口拿到 JSON 数据后，就可以使用例如 Vue.js、AngularJS、RectJS 等前端框架进行渲染。

## 单元小结

- MVC(Model-View-Controller)模式是软件工程中的一种软件架构模式，把软件系统分为三个基本部分：模型(Model)、视图(View)和控制器(Controller)。
- Spring MVC 是一种基于 Servlet 的技术。
- Spring MVC 框架是围绕 DispatcherServlet 工作的，这个是其核心类。
- Spring MVC 中控制器是开发的核心内容。

## 单元自测

1. 在 Spring MVC 中，以下描述错误的是(　　)。
   A. Spring MVC 中必须是实现了 Handler 接口的 JavaBean 才能成为请求处理器
   B. DispatcherServlet 是 Spring MVC 的前端 Servlet，和任何的 Servlet 一样，必须在 web. xml 中配置后，才能起作用
   C. 在 web. xml 中，根据 servlet-mapping 的 URL 不同，可以配置多个 DispatcherServlet
   D. ModelAndView 中的 view 是逻辑视图名，而非真正的视图对象
2. 有关 MVC 的处理过程，描述不正确的是(　　)。
   A. 首先控制器接收用户的请求，决定调用哪个模型来进行处理
   B. 模型处理用户的请求并返回数据
   C. 模型确定调用哪个视图进行数据展示
   D. 视图将模型返回的数据呈现给用户
3. 关于 Model 1 研发模式和 Model 2 开发模式的对比，下列说法错误的是(　　)。
   A. Model 2 的结构更加清晰
   B. Model 1 更利于小组分工合作
   C. Model 1 的组件更难于重用
   D. Model 1 开发出来的项目更难以维护
4. 下列关于 ModelAndView 的说法，错误的是(　　)。
   A. 控制器处理方法的返回值若为 ModelAndView，则既可以包含视图信息，也可以包含模型数据信息
   B. 控制器处理方法的返回值若为 ModelAndView，在处理方法的方法体内，除了通过 setViewName()或者 setView()设置视图，还必须通过 addobject()添加模型数据

C. Mode1AndView 的 addobject ()方法跟 addAllobjects ()方法区别：前者添加一个对象到 model 中，后者是添加一个 Map 对象

D. Mode1AndView. setViewName ("welcome") 中的 welcome 是逻辑视图名，并非真正的视图对象

5. 下列关于@RequestParam 绑定请求参数值的说法，正确的是(　　)。

A. Controller 方法入参时，必须使用@RequestParam 指定请求参数

B. @RequestParam 的参数 required，默认值为 false

C. 若有非必须的参数，必须使用@RequestParam 注解标注并设置 required=false

D. 使用@RequestParam 指定对应请求参数，并且 required=true，那么请求中若不加此参数，则会报请求错误

## 上机实战

### 上机目标

掌握 Spring MVC 开发流程。

### 上机练习

### 练习：完成用户登录功能

**【问题描述】**

实现用户登录功能，输入用户名或邮箱和密码并进行前后端校验。登录成功跳转到首页，并展示欢迎信息。

**【问题分析】**

本练习主要是为了巩固 Spring MVC 开发流程。

**【参考步骤】**

参照本单元的工程 springmvc-demo 自行完成。

# Spring MVC组件开发

## 课程目标

❖ 了解 ASP.NET MVC 开发模式

❖ 掌握请求参数的绑定方法

❖ 掌握重定向的方法

❖ 掌握拦截器的使用

**简介**

在上一单元中，我们熟悉了 Spring MVC 的开发流程，本单元将主要讨论 Spring MVC 的组件和开发流程，并对一些常用注解进行更为详细的讨论。

# 11.1 请求参数的绑定

使用控制器接收参数往往是 Spring MVC 开发业务逻辑的第一步，在上一单元中，我们了解了简单的参数传递，除了这种简单的参数传递外，实际开发中会有很多复杂的参数传递。比如现在流行的 Restful 风格，往往会将参数写入到路径中。JSON 格式数据的传递也很复杂。

## 11.1.1 普通参数的绑定

Spring MVC 绑定请求参数是自动实现的，只要满足下面的规则即可。

- 基本数据类型或 String 类型：要求参数名称必须和控制器中的方法的形参名称保持一致。(严格区分大小写)
- POJO 类型：要求表单中的参数名称和 POJO 类的属性名称保持一致，并且控制器方法的参数类型是 POJO 类型。
- 集合类型：要求集合类型的请求参数必须在 POJO 中。在表单中的请求参数名称要和 POJO 中的集合属性名称相同。如果是给 List 集合中的元素赋值，则使用下标；如果给 Map 集合中的元素赋值，则使用键值对。

下面针对这几种情况进行示例演示。

### 1. 基本类型或 String 类型作为参数

(1) 创建工程 springmvc-demo2，工程基础代码同上一单元中的 springmvc-demo 工程。在 WEB-INF/view 目录下创建 role.jsp 页面，代码如下。

```
<%@ page contentType="text/html;charset=UTF-8" %>
<!DOCTYPE html>
<html>
<head>
    <!-- 页面 meta -->
    <meta charset="utf-8">
    <meta http-equiv="X-UA-Compatible" content="IE=edge">
```

```
        <title>role</title>
        <link href="${pageContext.request.contextPath}/css/bootstrap.min.css" rel="stylesheet">
        <script src="${pageContext.request.contextPath}/js/jquery-3.2.1.min.js"></script>
        <script src="${pageContext.request.contextPath}/js/bootstrap.min.js"></script>
</head>
<body>
<div class="container">
        <div>
            <h3 class="text-center">角色添加</h3>
        </div>
        <form class="form-horizontal" action="${pageContext.request.contextPath}/params/add"
         method="post">
            <div class="form-group">
                <label for="roleName" class="col-sm-2 control-label">角色名称</label>
                <div class="col-sm-10">
                    <input type="text" class="form-control" name="roleName" id="roleName"
                     placeholder="请输入角色名称">
                </div>
            </div>
            <div class="form-group">
                <label for="roleDesp" class="col-sm-2 control-label">角色描述</label>
                <div class="col-sm-10">
                    <input type="text" class="form-control" name="roleDesp" id="roleDesp"
                     placeholder="请输入角色描述">
                </div>
            </div>
            <div class="form-group">
                <div class="col-sm-offset-2 col-sm-10">
                    <div class="radio">
                        <label class="radio-inline">
                            <input type="radio" name="enable" id="inlineRadio1" value="0"> 不可用
                        </label>
                        <label class="radio-inline">
                            <input type="radio" name="enable" id="inlineRadio2" value="1" checked> 可用
                        </label>
                    </div>
                </div>
            </div>
            <div class="form-group">
                <div class="col-sm-offset-2 col-sm-10">
                    <button type="submit" class="btn btn-default">添加</button>
                </div>
            </div>
        </form>
</div>
</body>
</html>
```

(2) 创建接收参数的控制器 ParamsController，代码如下所示。

```
package com.bailiban.controller;

import org.springframework.stereotype.Controller;
import org.springframework.web.bind.annotation.RequestMapping;
import org.springframework.web.bind.annotation.RequestMethod;

/**
 * @author zhulang
 * @Classname ParamsController
 * @Description 接收参数控制器
 * @Date 2020/8/20 10:49
 */
@Controller
@RequestMapping("/params")
public class ParamsController {
    @RequestMapping(value = "/add", method = RequestMethod.POST)
    public String addRole(String roleName, String roleDesp, Integer enable) {
        System.out.println(roleName);
        System.out.println(roleDesp);
        System.out.println(enable);
        return "success";
    }
}
```

这里主要为了测试参数是否绑定成功，因此并没有涉及调用业务层代码。

(3) 运行工程，在页面输入测试信息，如图 11-1 所示。

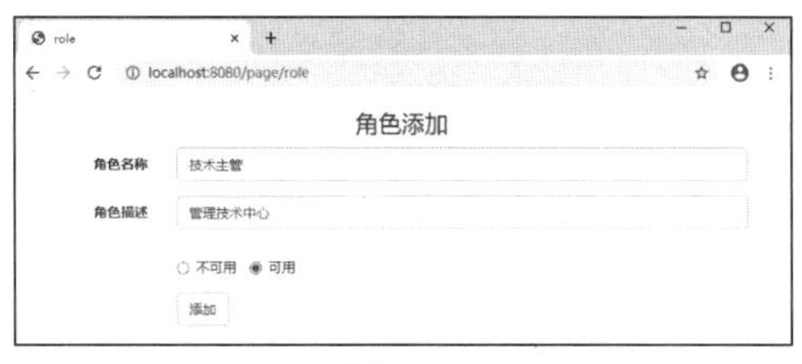


图 11-1

(4) 单击“提交”按钮，查看控制台输出的信息，如图 11-2 所示。

```
[INFO] Initializing Servlet 'dispatcher'
[INFO] Completed initialization in 3351 ms
八月 20, 2020 3:54:39 下午 org.apache.coyote.AbstractProtocol start
信息: Starting ProtocolHandler ["http-nio-8080"]
æŠ€æœ¯ä¸»ç®¡
ç®¡ç†æŠ€æœ¯ä¸å¿ƒ
1
```

图 11-2

这时会发现接收的参数是乱码，因为浏览器发送请求和解析 POST 数据的默认编码为“ISO-8859-1”，但是服务器响应的数据编码格式为 UTF-8。解决 POST 乱码的通用方法是在 web.xml 中加入字符编码过滤器，过滤所有请求，将编码格式设置为 utf-8。代码如下所示。

```
<!-- 过滤器解决 post 乱码-->
<filter>
    <filter-name>CharacterEncodingFilter</filter-name>
    <filter-class>org.springframework.web.filter.CharacterEncodingFilter</filter-class>
    <init-param>
        <param-name>encoding</param-name>
        <param-value>utf-8</param-value>
    </init-param>
</filter>
<filter-mapping>
    <filter-name>CharacterEncodingFilter</filter-name>
    <url-pattern>/*</url-pattern>
</filter-mapping>
```

(5) 重新运行工程进行测试，控制台输出的结果如图 11-3 所示。

```
[INFO] Initializing Servlet 'dispatcher'
[INFO] Completed initialization in 3359 ms
八月 20, 2020 4:20:05 下午 org.apache.coyote.AbstractProtocol start
信息: Starting ProtocolHandler ["http-nio-8080"]
技术主管
管理技术中心
1
```

图 11-3

由此可以发现对于基本数据类型或者 String 类型，在表单请求参数和控制器中方法参数名称一致的情况下，Spring MVC 可以自动完成数据的绑定。

### 2. POJO 类型

(1) 对上面控制器代码中的 addRole 方法进行修改，代码如下所示。

```
@RequestMapping(value = "/add", method = RequestMethod.POST)
public String addRole(Role role) {
    System.out.println(role);
    return "success";
}
```

(2) 同样的，重新运行工程，查看控制台输出的信息，如图 11-4 所示。

```
[INFO] Initializing Servlet 'dispatcher'
[INFO] Completed initialization in 3619 ms
八月 20, 2020 4:37:00 下午 org.apache.coyote.AbstractProtocol start
信息: Starting ProtocolHandler ["http-nio-8080"]
Role(id=null, roleName=技术主管, roleDesp=管理技术中心, enable=1)
```

图 11-4

由此可以发现对于 POJO 类型，在表单请求参数和控制器中方法 POJO 参数属性名称一致的情况下，Spring MVC 可以自动完成数据的绑定。

### 3. 集合类型

对于集合类型，Spring MVC 也是支持自动绑定的。但由于集合类型的参数绑定，实际开发中更多的是前端采用传递 JSON 格式的数据的方式，这里就不作演示。

## 11.1.2　使用注解@RequestParam 绑定参数

前面我们提到对于基本数据类型或 String 类型，要求请求参数名称和控制器中方法参数名称一致，Spring MVC 能够完成自动绑定参数。如果参数名称不一致，获取参数就会失败，此时可以使用注解@RequestParam 来解决这个问题。修改控制器代码为如下所示。

```
@RequestMapping(value = "/add", method = RequestMethod.POST)
public String addRole(@RequestParam("roleName") String name, @RequestParam("roleDesp") String
                    desp, Integer enable) {
    System.out.println(name);
    System.out.println(desp);
    System.out.println(enable);
    return "success";
}
```

这样就可以将请求参数 roleName 的值绑定到参数 name 上，请求参数 roleDesp 的值绑定到参数 desp 上。

## 11.1.3　使用注解@PathVariable 绑定 URL 参数

有时候可能会将参数通过 URL 的形式进行传递，例如分页信息、通过 id 查询角色信息，这符合 Restful 风格，Spring MVC 支持这种方式，并提供了注解@PathVariable 和@RequestMapping 来协同完成。如下代码所示。

```
@RequestMapping("/{page}/{size}")
public String pageRole(@PathVariable("page") Integer page, @PathVariable("size") Integer size) {
    System.out.println(page);
    System.out.println(size);
    return "success";
}
```

上面代码中的{page}和{size}代表处理器需要接收两个由 URL 组成的参数，且参数名称为 page 和 size，在方法中的@PathVariable("page")和@PathVariable("size") 表示将获取在@RequestMapping 中定义的名称为 page 和 size 的参数。运行工程，在浏览器输入 http://localhost:8080/params/1/10，查看控制台的信息，如图 11-5 所示。

```
[INFO] Completed initialization in 3269 ms
八月 21, 2020 9:11:42 上午 org.apache.coyote.AbstractProtocol start
信息: Starting ProtocolHandler ["http-nio-8080"]
1
10
```

图 11-5

从图中可以发现，成功获取了请求路径中的 page 和 size 的值，这样就可以通过@PathVariable 注解获取各类参数。另外@PathVariable 允许对应的参数为空。

## 11.1.4 使用注解@RequestBody 绑定 JSON 数据

有时参数的传递还需要更多的参数，是一种集合，此时前端向后台传递参数更多的是采用 JSON 格式，例如要查询员工信息，查询参数有员工姓名、生日、性别和地址。其实体类代码如下所示。

```
package com.bailiban.entity;

import com.fasterxml.jackson.annotation.JsonFormat;
import lombok.Data;

import java.io.Serializable;
import java.util.Date;
import java.util.List;

/**
 * (Emp)实体类
 *
 * @author zhulang
 * @since 2020-08-21 10:49:38
 */
@Data
public class Emp implements Serializable {
    private static final long serialVersionUID = 531172534075955547L;

    private Integer id;
    /**
     * 员工姓名
     */
    private String username;
    /**
     * 生日
     */
    @JsonFormat(pattern = "yyyy-MM-dd", timezone = "GMT+8")
    private Date birthday;
    /**
```

```
        * 性别
        */
      private String sex;
      /**
        * 地址
        */
      private String address;

      private List<Role> roles;
}
```

上面代码是 Emp 员工实体类，一个员工可以拥有多个角色，另外在 birthday 属性上面我们使用了一个@JsonFormat 注解，用于将数据库查询出来的日期类型进行格式化为指定的时间格式。

下面来看页面代码 emp.html，如下所示。

```
<!DOCTYPE html>
<html lang="en">
<head>
      <meta charset="UTF-8">
      <title>员工管理</title>
      <link type="text/css" rel="stylesheet" href="css/bootstrap.min.css"/>
      <script src="js/jquery-3.2.1.min.js"></script>
      <script src="js/bootstrap.min.js"></script>
</head>
<body>
<div class="container">
      <table style="width: 500px;" class="table table-striped table-hover table-bordered">
            <thead>
            <tr>
                  <th>ID</th>
                  <th>姓名</th>
                  <th>生日</th>
                  <th>性别</th>
                  <th>地址</th>
            </tr>
            </thead>
            <tbody id="content">

            </tbody>
      </table>
</div>
</body>
<script>
      $.ajax({
            url: "/emp/search",//请求后台 API
            type: "POST",
```

```
                dataType: "json",
                success: function(data){
                    $.each(data, function(index, value) {//循环后台返回的数据
                        let content = "<tr>" +
                            "<td>"+value['id']+"</td>"+
                            "<td>"+value['username']+"</td>"+
                            "<td>"+value['birthday']+"</td>"+
                            "<td>"+value['sex']+"</td>"+
                            "<td>"+value['address']+"</td>"+
                            "</tr>";
                        $("#content").append(content);//追加内容
                    });
                },
                error:function(err){
                    console.log(err);
                }
        });
    </script>
    </html>
```

为了让大家对页面有更多的认识，不再局限于 jsp，这里使用的是 html，里面用到了一点 jQuery 和 JavaScript 的语法。

为了防止 emp.html 页面被拦截器拦截，在 springmvc.xml 中，我们加入了如下配置。

```
<mvc:resources location="/" mapping="*.html"/>
```

同时将 emp.html 页面放在 webapp 目录下，这样就可以直接访问到 emp.html。

启动服务器，运行成功后，在浏览器地址栏中输入地址 http://localhost:8080/emp.html。页面截图如图 11-6 所示。

| ID | 姓名 | 生日 | 性别 | 地址 |
|---|---|---|---|---|
| 1 | 维亚 | 1999-10-25 | 女 | 浙江杭州滨江新园区 |
| 2 | 阿伦 | 1987-10-25 | 男 | 上海浦东新区 |
| 3 | 玥玥 | 2013-01-15 | 女 | 广东深圳南山区 |
| 4 | 张良 | 2009-10-25 | 男 | 湖北武汉东湖高新 |
| 5 | 宋宝 | 2020-05-13 | 男 | 湖北武汉洪山区 |

图 11-6

## 11.2 转发和重定向

转发和重定向是在学习 Servlet 时就接触过的内容，学习 Spring MVC 后会发现可以使

用更加简便的方式来实现。这里只讨论转发和重定向比较常用的业务场景。

## 11.2.1　转发

在控制器中返回 String 类型的值时，默认就是请求转发。代码如下所示。

```
@RequestMapping("/forward")
public String testForward() {
    System.out.println("testForward 方法执行了……");
    return "success";
}
```

也可以写成如下形式。

```
@RequestMapping("/forward")
public String testForward() {
    System.out.println("testForward 方法执行了……");
    return "forward:/WEB-INF/view/success.jsp";
}
```

需要注意的是，使用 forward:，路径必须写成实际视图 url，不能写逻辑视图。转发更多的是用于页面的跳转，虽然也可以进行控制器方法间的跳转，但更多的会使用重定向来实现。

## 11.2.2　重定向

在很多时候，我们会碰到新增、修改或删除等操作后重新查询返回展示页面的情况。这时使用重定向是最合适的方式之一。例如在 emp.html 页面中添加 Emp 功能，添加完成后，重新展示所有员工信息。控制器代码如下所示。

```
package com.bailiban.controller;

import com.bailiban.entity.Emp;
import com.bailiban.service.EmpService;
import org.springframework.beans.factory.annotation.Autowired;
import org.springframework.stereotype.Controller;
import org.springframework.web.bind.annotation.RequestBody;
import org.springframework.web.bind.annotation.RequestMapping;
import org.springframework.web.bind.annotation.ResponseBody;

import java.util.List;

/**
 * @author zhulang
```

```
 * @Classname EmpController
 * @Description 员工类控制器
 * @Date 2020/8/24 10:49
 */
@Controller
@RequestMapping("/emp")
public class EmpController {
    @Autowired
    private EmpService empService;

    @RequestMapping("/search")
    @ResponseBody
    public List<Emp> search(@RequestBody(required = false) Emp emp) {
        System.out.println("查询");
        return empService.findAll(emp);
    }

    @RequestMapping("/delIds")
    @ResponseBody
    public int deleteEmps(@RequestBody Integer[] ids) {
        return empService.deleteByIds(ids);
    }

    @RequestMapping("/add")
    public String addEmp(@RequestBody Emp emp) {
        empService.addEmp(emp);
        return "redirect:/emp/search";
    }
}
```

控制器中方法 addEmp 完成添加操作后，会跳转到控制器中路径为/emp/search 对应的 search 方法中，进而完成查询操作。同样的重定向也可以用于页面跳转，需要注意的是当重定向到页面时，页面不能放在 WEB-INF 目录下，否则会找不到，此时只能使用请求转发。

## 11.3 保存并获取属性参数

在 JavaEE 的基础学习中，有时候我们会暂存数据到 HTTP 的 request 对象或者 Session 对象中，现在 Spring MVC 提供了三个注解来更好地完成这些任务，分别是@RequestAttribute、@SessionAttribute 和@SessionAttributes，对它们的作用解释如下。

- @RequestAttribute：获取 HTTP 的请求(Request)对象属性值，用来传递给控制器的参数。
- @SessionAttribute：在 HTTP 的会话(Session)对象属性值中，用来传递给控制器的参数。
- @SessionAttributes：可以将数据模型中的数据存储到 Session 中。

下面一一对它们进行讨论。

## 11.3.1　注解@RequestAttribute

@RequestAttribute 的主要作用是从 HTTP 的 request 对象中取出请求属性，我们在 webapp 下新建一个 attribute.jsp 文件，内容如下。

```
<%@ page contentType="text/html;charset=UTF-8" language="java" %>
<!DOCTYPE html>
<html>
<head>
    <!-- 页面 meta -->
    <meta charset="utf-8">
    <meta http-equiv="X-UA-Compatible" content="IE=edge">
    <title>attribute</title>
</head>
<body>
<%
    // 设置请求属性
    request.setAttribute("id", 1);
    // 转发给控制器
    request.getRequestDispatcher("/attribute/requestAttribute").forward(request, response);
%>
</body>
</html>
```

上面的代码中，在 request 域中存储了一个 id，值为 1，下面新建一个 AttributeController，代码如下所示。

```
package com.bailiban.controller;

import org.springframework.stereotype.Controller;
import org.springframework.web.bind.annotation.RequestAttribute;
import org.springframework.web.bind.annotation.RequestMapping;

/**
 * @author zhulang
 * @Classname AttributeController
 * @Description attribute 控制器
```

```
 * @Date 2020/8/28 10:03
 */
@Controller
@RequestMapping("/attribute")
public class AttributeController {
    @RequestMapping("/requestAttribute")
    public String reqAttr(@RequestAttribute("id") Integer id) {
        System.out.println(id);
        return "success";
    }
}
```

启动项目，在浏览器地址栏中输入地址：localhost:8080/attribute.jsp，可以看到控制台中输出的结果如图 11-7 所示。

```
信息: Initializing Spring DispatcherServlet 'dispatcher'
[INFO] Initializing Servlet 'dispatcher'
[INFO] Completed initialization in 2813 ms
八月 28, 2020 10:07:59 上午 org.apache.coyote.AbstractProtocol start
信息: Starting ProtocolHandler ["http-nio-8080"]
1
```

图 11-7

需要注意的是使用@RequestAttribute 注解，默认参数的值不能为空，否则会抛出异常。如果允许为空，需要配置 required=false。

## 11.3.2 注解@SessionAttribute 和@SessionAttributes

这两个注解看着相似，都是用来获取 Session 中存储的数据，但是@SessionAttribute 注解只能作用在方法的参数中，而@SessionAttributes 是作用在控制器上。

修改 attribute.jsp 代码，如下所示。

```
<%@ page contentType="text/html;charset=UTF-8" language="java" %>
<!DOCTYPE html>
<html>
<head>
    <!-- 页面 meta -->
    <meta charset="utf-8">
    <meta http-equiv="X-UA-Compatible" content="IE=edge">
    <title>attribute</title>
</head>
<body>
<%
    // 设置请求属性
    request.setAttribute("id", 1);
    session.setAttribute("name", "百里半");
    // 转发给控制器
```

```
        request.getRequestDispatcher("/attribute/sessionAttribute").forward(request, response);
    %>
    </body>
    </html>
```

上面的代码中，在 Session 域中存储了一个 name，值为“百里半”，下面在 AttributeController 中新增一个方法，代码如下所示。

```
@RequestMapping("/sessionAttribute")
public String sessionAttr(@SessionAttribute("name") String name) {
    System.out.println(name);
    return "success";
}
```

重新启动项目，在浏览器地址栏输入地址：localhost:8080/attribute.jsp，查看控制台输出的结果，如图 11-8 所示。

```
[INFO] Initializing Servlet 'dispatcher'
[INFO] Completed initialization in 2960 ms
八月 28, 2020 10:32:30 上午 org.apache.coyote.AbstractProtocol start
信息: Starting ProtocolHandler ["http-nio-8080"]
百里半
```

图 11-8

同理@SessionAttribute 默认参数的值不能为空，否则会抛出异常。如果允许为空，需要配置 required=false。

下面再来测试@SessionAttributes，控制器代码如下。

```
@Controller
@RequestMapping("/attribute")
@SessionAttributes(names = {"id", "name"})
public class AttributeController {
    @RequestMapping("/requestAttribute")
    public String reqAttr(@RequestAttribute("id") Integer id) {
        System.out.println(id);
        return "success";
    }

    @RequestMapping("/sessionAttribute")
    public String sessionAttr(@SessionAttribute("name") String name) {
        System.out.println(name);
        return "success";
    }

    @RequestMapping("/sessionAttributes")
    public String sessionAttrs(Model model) {
        model.addAttribute("id", 2);
        model.addAttribute("name", "bailiban");
```

```
            return "success";
        }
}
```

在控制器方法 sessionAttrs 中向 Model 对象添加了对应的键值对。Model 中存入的键值对是 request 级别的，但是因为在 AttributeController 控制器上添加了注解 @SessionAttributes，设置了属性名称 id 和 name。此时 model 中的 id 和 name 会同步存储到 Session 域中。

修改 success.jsp 代码，取出 request 和 session 域中的值，代码如下所示。

```
<%@ page contentType="text/html;charset=UTF-8" %>
<!DOCTYPE html>
<html>
<head>
    <!-- 页面 meta -->
    <meta charset="utf-8">
    <meta http-equiv="X-UA-Compatible" content="IE=edge">
    <title>success</title>
</head>
<body>
<h1>SUCCESS</h1>
<h5>session 域</h5>
id: ${sessionScope.id}
<br>
name: ${sessionScope.name}
</body>
</html>
```

重新启动项目，在浏览器地址栏输入：localhost:8080/attribute/sessionAttributes，查看浏览器响应结果，如图 11-9 所示。

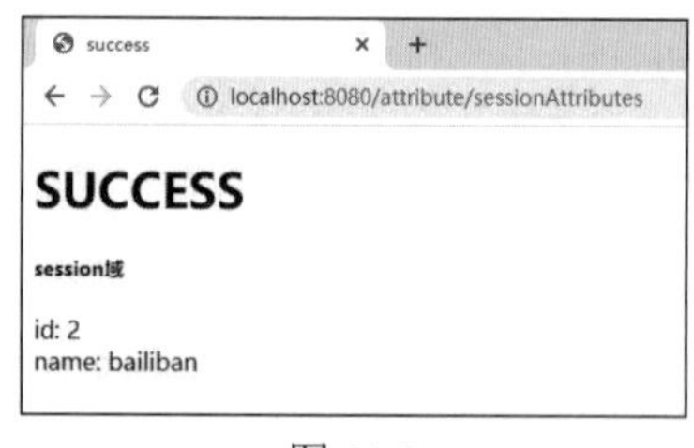


图 11-9

此时 Session 域中获取到了 id 和 name 的值。

如果要清空 Session，可以使用 SessionStatus 的 setComplete()方法，代码如下所示。

```
@RequestMapping("/cleanSession")
public String sessionStatus(SessionStatus status){
    status.setComplete();
    return "success";
```

```
}
```

再次运行工程，先在浏览器地址栏输入：localhost:8080/attribute/sessionAttributes，然后再输入 localhost:8080/attribute/cleanSession，可以看到页面响应结果，如图 11-10 所示。

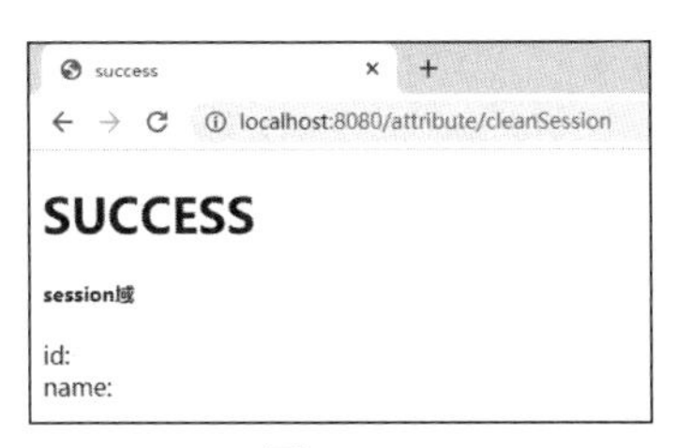


图 11-10

### 11.3.3　注解@CookieValue 和@RequestHeader

这两个注解分别是从 Cookie 和 HTTP 请求头获取对应的请求信息，并不是很常用，下面给出简单的实例代码，如下所示。

```
@RequestMapping("/cookieAndHeader")
public String cookieAndHeader(@CookieValue(value = "JSESSIONID", required = false) String cookie,
                              @RequestHeader(value = "Accept-Language", required = false) String header) {
    System.out.println("cookie:" + cookie);
    System.out.println("header:" + header);
    return "success";
}
```

重新运行工程，在浏览器地址栏输入：localhost:8080/attribute/cookieAndHeader，查看控制台显示的结果，如图 11-11 所示。

```
[INFO] Completed initialization in 6382 ms
八月 28, 2020 2:38:21 下午 org.apache.coyote.AbstractProtocol start
信息: Starting ProtocolHandler ["http-nio-8080"]
cookie:AC7B4BC7F824AD6826D560AF45D8917B
header:zh-CN,zh;q=0.9
```

图 11-11

## 11.4　拦截器

Spring MVC 的处理器拦截器类似于 Servlet 开发中的过滤器 Filter，用于对处理器进行预处理和后处理。用户还可以自定义一些拦截器来实现特定的功能。

在谈到拦截器，我们还要提到拦截器链(Interceptor Chain)这个词，它是将拦截器按照一定的顺序联结成一条链。在访问被拦截的方法或者字段时，拦截器链中的拦截器就会按其定义的顺序被调用。下面一起来讨论拦截器的功能和流程。

### 11.4.1 拦截器的定义

Spring MVC 拦截器的实现一般有两种方式：第一种方式是要定义的 Interceptor 类要实现 Spring 的 HandlerInterceptor 接口；第二种方式是继承实现了 HandlerInterceptor 接口的类，比如 Spring 已经提供的实现了 HandlerInterceptor 接口的抽象类 HandlerInterceptorAdapter。

下面来看 HandlerInterceptor 接口的源码，如下所示。

```
package org.springframework.web.servlet;

import javax.servlet.http.HttpServletRequest;
import javax.servlet.http.HttpServletResponse;
import org.springframework.lang.Nullable;

public interface HandlerInterceptor {
    default boolean preHandle(HttpServletRequest request, HttpServletResponse response,
                              Object handler) throws Exception {
        return true;
    }

    default void postHandle(HttpServletRequest request, HttpServletResponse response, Object handler,
                            @Nullable ModelAndView modelAndView) throws Exception {
    }

    default void afterCompletion(HttpServletRequest request, HttpServletResponse response,
                                 Object handler, @Nullable Exception ex) throws Exception {
    }
}
```

接口中定义了三个方法，其含义解释如下。

- preHandle：在处理器之前执行的前置方法，如果该方法返回 true，则请求继续向下进行，否则请求不会继续向下进行，处理器也不会被调用。
- postHandle：在处理器之后执行的后置方法。
- afterCompletion：只要该拦截器中的 preHandle 方法返回 true，该方法就会在渲染视图后被调用。

### 11.4.2 拦截器的执行流程

学习拦截器时，一定要了解其执行流程，这样才能更好地应用它，如图 11-12 所示。

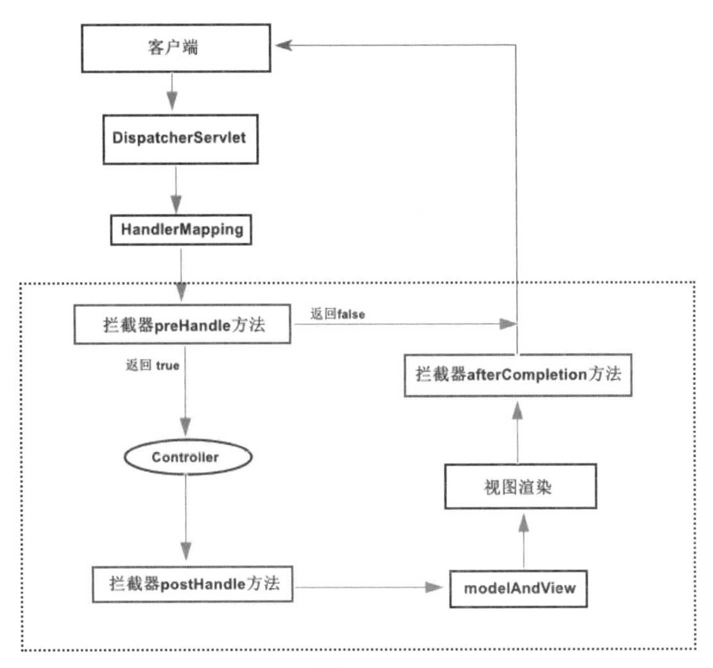


图 11-12

从图 11-12 中可以看出拦截器中的前置方法返回值决定了后续的流程是否执行，同时需要注意的是后置方法 postHandle 是在 handler 方法(Controller)成功执行之后、dispatcher 渲染视图之前执行。因此如果 handler 方法抛出异常，则 postHandle 也是不会执行的。但是 afterCompletion 只要前置方法返回 true，后面无论是否抛出异常，都会在视图渲染结束后执行。

## 11.4.3　开发拦截器

前面提到过实现拦截器可以实现 HandlerInterceptor 接口或者继承实现了 HandlerInterceptor 接口的类。这里直接编写一个 LoginInterceptor 去实现 HandlerInterceptor 接口，代码如下所示。

```
package com.bailiban.interceptor;

import org.springframework.web.servlet.HandlerInterceptor;
import org.springframework.web.servlet.ModelAndView;

import javax.servlet.http.HttpServletRequest;
import javax.servlet.http.HttpServletResponse;

/**
 * @author zhulang
 * @Classname LoginInterceptor
```

```
     * @Description 登录拦截器
     * @Date 2020/8/28 16:53
     */
    public class LoginInterceptor implements HandlerInterceptor {
        @Override
        public boolean preHandle(HttpServletRequest request, HttpServletResponse response,
          Object handler) throws Exception {
            System.out.println("---前置方法执行---");
            return true;
        }

        @Override
        public void postHandle(HttpServletRequest request, HttpServletResponse response, Object handler,
                                ModelAndView modelAndView) throws Exception {
            System.out.println("---后置方法执行---");
        }

        @Override
        public void afterCompletion(HttpServletRequest request, HttpServletResponse response, Object
handler, Exception ex) throws Exception {
            System.out.println("---完成方法执行---");
        }
    }
```

此外还需要在 Spring MVC 配置文件中进行配置，加入如下配置代码。

```
<!--配置拦截器-->
<mvc:interceptors>
    <mvc:interceptor>
        <mvc:mapping path="/**"/>
        <bean id="loginInterceptor" class="com.bailiban.interceptor.LoginInterceptor"/>
    </mvc:interceptor>
</mvc:interceptors>
```

元素<mvc:interceptors>里面可以配置多个拦截器，每个拦截器拦截什么样的请求通过<mvc:mapping>中的 path 属性进行设置，上面代码中的 path="/**"表示拦截所有请求。这样就完成了一个简单的拦截器的开发。

### 11.4.4 多个拦截器执行的顺序

关于多个拦截器的执行顺序，下面用代码进行演示。先编写三个拦截器，这里就直接复制两个 LoginInterceptor 并分别命名为 LoginInterceptor2、LoginInterceptor3。在 Spring MVC 中配置拦截器的代码如下所示。

```
<!-- 配置拦截器-->
<mvc:interceptors>
```

```
        <mvc:interceptor>
            <mvc:mapping path="/**"/>
            <bean id="loginInterceptor" class="com.bailiban.interceptor.LoginInterceptor"/>
        </mvc:interceptor>
        <mvc:interceptor>
            <mvc:mapping path="/**"/>
            <bean id="loginInterceptor2" class="com.bailiban.interceptor.LoginInterceptor2"/>
        </mvc:interceptor>
        <mvc:interceptor>
            <mvc:mapping path="/**"/>
            <bean id="loginInterceptor3" class="com.bailiban.interceptor.LoginInterceptor3"/>
        </mvc:interceptor>
    </mvc:interceptors>
```

运行项目，在浏览器地址栏输入一个能进入控制器的路径，如：localhost:8080/attribute/cookieAndHeader，查看控制台输出的信息，如图 11-13 所示。

```
信息: Initializing Spring DispatcherServlet 'dispatcher'
[INFO] Initializing Servlet 'dispatcher'
[INFO] Completed initialization in 3544 ms
八月 28, 2020 5:51:33 下午 org.apache.coyote.AbstractProtocol start
信息: Starting ProtocolHandler ["http-nio-8080"]
---拦截器1---前置方法执行---
---拦截器2---前置方法执行---
---拦截器3---前置方法执行---
cookie:E65CBA435D934FBBD377011454D47C74
header:zh-CN,zh;q=0.9
---拦截器3---后置方法执行---
---拦截器2---后置方法执行---
---拦截器1---后置方法执行---
---拦截器3---完成方法执行---
---拦截器2---完成方法执行---
---拦截器1---完成方法执行---
```

图 11-13

从图中可以看出，正常情况下，Spring 会先从第一个拦截器开始进入前置方法，然后按照配置中的拦截器顺序依次执行第二个、第三个的前置方法。后置方法和完成方法则刚好是与之相反的顺序。

需要注意的是，以上是在正常的情况下，如果有些前置方法返回 false，那么结果就又不一样了。下面将 LoginInterceptor2 中的前置方法返回值修改为 false 进行测试。输出的结果如图 11-14 所示。

```
[INFO] Completed initialization in 3794 ms
八月 28, 2020 6:02:12 下午 org.apache.coyote.AbstractProtocol start
信息: Starting ProtocolHandler ["http-nio-8080"]
---拦截器1---前置方法执行---
---拦截器2---前置方法执行---
---拦截器1---完成方法执行---
```

图 11-14

从输出结果可以看出，当其中一个前置方法返回 false，那么后面的前置方法、控制器方法、后置方法都不会执行，只有对应返回 true 的拦截器的完成方法(afterCompletion)才会逆序执行。

# 11.5 文件上传

在互联网项目中，文件上传是一个很常见的业务功能，例如上传图像、图片、证件、相关文件等。Spring MVC 为文件上传提供了良好的支持。Spring MVC 是通过 MultipartResolver(Multipart 解析器)处理的，它是一个接口，有以下两个实现类。

- CommonsMultipartResolver：使用 Jakarta Commons FileUpload 解析 multipart 请求，依赖于第三方包。
- StandardServletMultipartResolver：依赖于 Servlet 3.0 对 multipart 请求的支持(始于 Spring 3.1)，它不用依赖第三方包。

一般来说，StandardServletMultipartResolver 将会是优先选择的方案，它使用了 Servlet 所提供的原生功能支持，并不需要依赖任何第三方组件。这里提到的两种方式我们都会进行演示。

## 11.5.1 MultipartResolver 概述

在上面我们提到 Spring MVC 是通过 MultipartResolver 解析器进行上传处理的，因此需要注入 MultipartResolver 到 Spring 容器中，可以选择使用 Java 配置或者使用 XML 配置方式。这里使用 XML 配置。在 applicationContext.xml 中加入如下代码。

```
<bean name="multipartResolver"
 class="org.springframework.web.multipart.support.StandardServletMultipartResolver"/>
```

上面配置中 name 的值是 Spring 约定好的 multipartResolver。除了配置 Bean 外，有时候还需要对上传作一些配置，比如限制单个文件上传的大小、总文件上传的大小等。下面在 web.xml 中进行如下配置。

```
<!--配置前端控制器-->
<servlet>
    <!--名称-->
    <servlet-name>dispatcher</servlet-name>
    <!-- Servlet 类-->
    <servlet-class>org.springframework.web.servlet.DispatcherServlet</servlet-class>
    <init-param>
        <!--Spring MVC 配置参数文件的位置-->
        <param-name>contextConfigLocation</param-name>
        <!--(可不配置)默认名称为 ServletName-servlet.xml,即:dispatcher-servlet.xml-->
        <param-value>classpath*:spring/springmvc.xml</param-value>
    </init-param>
    <!--启动顺序，数字越小，启动越早，非 0 正整数-->
```

```
        <load-on-startup>1</load-on-startup>
        <!--文件上传的路径和大小-->
        <multipart-config>
            <!--上传文件的临时目录-->
            <location>/</location>
            <!--单个数据最大 5M-->
            <max-file-size>5242880</max-file-size>
            <!--总上传数据最大 10M-->
            <max-request-size>10485760</max-request-size>
            <!--文件缓存的临界点-->
            <file-size-threshold>0</file-size-threshold>
        </multipart-config>
    </servlet>
```

在<servlet>中加入<multipart-config>配置元素，对 MultipartResolver 进行初始化。

以上是对 StandardServletMultipartResolver 的配置过程，针对 CommonsMultipartResolver 的配置也比较简单，需要加入以下第三方依赖包。

```
    <dependency>
        <groupId>commons-fileupload</groupId>
        <artifactId>commons-fileupload</artifactId>
        <version>1.4</version>
    </dependency>
```

同理，也需要将 CommonsMultipartResolver 注入 Spring 容器中，在 applicationContext.xml 中加入如下配置。

```
    <bean name="multipartResolver"
     class="org.springframework.web.multipart.commons.CommonsMultipartResolver">
        <!--单个文件上传最大值-->
        <property name="maxUploadSizePerFile">
            <value>5242880</value>
        </property>
        <!--总文件上传最大值-->
        <property name="maxUploadSize">
            <value>10485760</value>
        </property>
    <!--上传文件的临时目录-->
        <property name="uploadTempDir">
            <value>/</value>
        </property>
    </bean>
```

这里 Bean 的 name 依旧是 Spring 约定的 MultipartResolver，是不能修改的。

## 11.5.2 文件上传实例

配置好 MultipartResolver 之后，接下来进行具体的代码实例演示，上传用户图像并回显。首先直接在项目中加入 imgUpload.html 文件，放在 imgUpload.html 中，如下所示。

```
<!DOCTYPE html>
<html lang="en">
<head>
    <meta charset="UTF-8">
    <title>上传图片并预览</title>
    <script src="js/jquery-3.2.1.min.js"></script>
</head>
<body>
    <div class="form-group">
        <label>图片</label>
        <img src="" alt="" id="img" />
        <input type="file" name="file" id="uploadImage" onchange="postData()">
    </div>
</body>
<script type="text/javascript">
    function postData() {
        var formData = new FormData();
        formData.append("file", $("#uploadImage")[0].files[0]);
        $.ajax({
            url: "file/upload",
            type: "post",
            data: formData,
            processData: false, // 告诉 jQuery 不要去处理发送的数据
            contentType: false, // 告诉 jQuery 不要去设置 Content-Type 请求头
            dataType: 'text',
            success: function(data) {
                $("#img").attr("src", "/upload/"+data);
            },
            error: function(data) {
            }
        });
    }
    // 赋值变量
    var basePath = getContextPath();
    // 获取项目路径
    function getContextPath() {
        var pathName = window.document.location.pathname;
        var projectName = pathName.substring(0, pathName.substr(1).indexOf(
            '/') + 1);
        return projectName;
    }
```

```
</script>
</html>
```

在 input 加上 postData 方法，利用 JavaScript 的 FormData 函数上传图片。

新建一个控制器 UploadController，代码如下。

```
package com.bailiban.controller;

import org.springframework.stereotype.Controller;
import org.springframework.web.bind.annotation.RequestMapping;
import org.springframework.web.bind.annotation.ResponseBody;
import org.springframework.web.multipart.MultipartFile;

import java.io.File;
import java.util.UUID;

/**
 * @author zhulang
 * @Classname UploadController
 * @Description  文件上传控制器
 * @Date 2020/9/1 9:50
 */
@Controller
@RequestMapping("/file")
public class UploadController {
    @RequestMapping("/upload")
    @ResponseBody
    public String upload(MultipartFile file) {
        if (file.isEmpty()) {
            return "上传文件为空";
        }
        // 获取文件全名 a.py
        String fileName = file.getOriginalFilename();
        // 文件上传路径
        String templatePath ="D:/Server/Tomcat9/webapps/upload";
    System.out.println("文件路径:" + templatePath);
        // 加上随机数，防止文件名重复，文件被覆盖
        fileName = UUID.randomUUID() + fileName;
        File dest0 = new File(templatePath);
        File dest = new File(dest0, fileName);
        // 文件上传
        try {
            // 检测是否存在目录
            if (!dest0.exists()) {
                dest0.mkdirs();
            }
// 保存文件
```

```
                file.transferTo(dest);
                return "上传成功";
            } catch (Exception e) {
                e.printStackTrace();
                return "上传失败";
            }
        }
    }
```

upload 方法中的参数 MultipartFile 是一个 Spring MVC 提供的类，可以接收到上传的文件。

启动项目，在浏览器中输入地址：localhost:8080/emp.html，单击上传窗口，选择图片即可成功上传并回显，如图 11-15 所示。

图 11-15

## 单元小结

- 对于基本数据类型或 String 类型，当请求参数名称和控制器中方法参数名称一致时，Spring MVC 能够完成自动绑定参数。
- 注解@PathVariable 和@RequestMapping 可以协同完成 Restful 风格的请求。
- 重定向是指浏览器向服务器发送一个请求并收到响应后再次向一个新地址发出请求，浏览器的地址栏会发生变化。
- 转发是服务器收到请求后为了完成响应跳转到一个新的地址，是服务器内部的跳转，浏览器的地址栏不会发生变化。
- Spring MVC 中的拦截器主要用于拦截用户请求并作相应的处理。常见的业务场景有进行权限验证、记录请求信息的日志、判断用户是否登录等。

- Spring MVC 提供两种上传方式配置，不论是哪种，其 Bean 的名字都只能为 multipartResolver。

## 单元自测

1. 下列关于 Spring MVC 的注解描述，错误的是(　　)。
   A. @ResponseBody 注解实现将 controller 方法返回对象转化为 JSON 响应给客户
   B. @RequestMapping 注解用于请求 url 映射
   C. @RequestBody 注解实现接收 http 请求的 JSON 数据，将 JSON 数据转换为 java 对象
   D. @PathVariable 注解可以独自完成路径中的参数映射
2. Spring MVC 的控制器默认模式是(　　)。
   A. 单例
   B. 原型
   C. 无
   D. 单例或原型
3. 以下有关 Spring MVC 配置文件中拦截器的配置说法，错误的是(　　)。
   A. 要使用 Spring MVC 中的拦截器，要先自定义拦截器，还需要在配置文件中进行配置
   B. < mvc:interceptors>元素用于配置一组拦截器，其子元素< bean>中定义的是指定路径的拦截器
   C. < mvc:interceptors>元素中可以同时配置多个< mvc:interceptor>子元素
   D. < mvc:exclude-mapping>元素用于配置不需要拦截的路径请求
4. 下列关于拦截器的执行流程说法，错误的是(　　)。
   A. 程序首先会执行拦截器类中的 preHandle()方法
   B. 如果 preHandle()方法的返回值为 true，则程序会继续向下执行处理器中的方法，否则将不再向下执行
   C. 在业务处理器(即控制器 Controller 类)处理完请求后，会执行 preHandle()方法
   D. 在 DispatcherServlet 处理完请求后，才会执行 afterCompletion()方法

5. 下面属于 Restful 风格请求的是(　　)。

A. http://…/queryItems?id=1

B. http://…/queryItems?id=1&name=zhangsan

C. http://…/items/1

D. http://…/queryitems/1

## 上机实战

### 上机目标

- 掌握 Spring MVC 中各种注解的使用。
- 掌握 Spring MVC 拦截器的使用。

### 上机练习

### 练习：使用 Spring MVC 拦截器实现登录功能

**【问题描述】**

只有登录成功的用户才能访问系统其他页面，否则无法访问，会直接调转到登录页。

**【问题分析】**

(1) 使用拦截器拦截用户的所有请求，除了登录之外。

(2) 在用户登录成功之后会把用户信息存储到session中，在拦截器的前置方法中判断session中是否有用户信息，从而决定是否放行。

**【参考步骤】**

(1) 在webapp/view目录下创建login.jsp登录页面。

```
<%@ page contentType="text/html;charset=UTF-8" %>
<!DOCTYPE html>
<html>
<head>
    <!-- 页面 meta -->
    <meta charset="utf-8">
    <meta http-equiv="X-UA-Compatible" content="IE=edge">
    <title>登录</title>
    <link href="${pageContext.request.contextPath}/css/bootstrap.min.css" rel="stylesheet">
    <script src="${pageContext.request.contextPath}/js/jquery-3.2.1.min.js"></script>
    <script src="${pageContext.request.contextPath}/js/bootstrap.min.js"></script>
</head>
<body>
<div class="container">
```

```
        <form action="${pageContext.request.contextPath}/emp/login" method="post">
            <div class="form-group col-md-6 col-md-offset-3">
                <label for="username">用户名</label>
                <input type="text" class="form-control" name="username" id="username" placeholder="用户名">
            </div>
            <div class="form-group col-md-6 col-md-offset-3">
                <label for="password">密码</label>
                <input type="password" class="form-control" name="password" id="password"
                placeholder="密码">
            </div>
            <div class="col-md-6 col-md-offset-3">
                <button type="submit" class="btn btn-default">登录</button>
            </div>
        </form>
    </div>
    </body>
    </html>
```

(2) 编写控制器代码，可以直接在 EmpController 下新增以下方法。这里假定只有用户名和密码分别为 root 和 admin 时方可成功登录。

```
@RequestMapping(value = "/login",method = RequestMethod.GET)
public String login() {
    return "login";
}

@RequestMapping(value = "/login",method = RequestMethod.POST)
public String doLogin(HttpSession session, String username, String password) {
    if (username.equals("root") && password.equals("admin")) {
        session.setAttribute("user", "username");
        return "success";
    } else {
        return "login";
    }
}
```

(3) 编写 LoginInterceptor 拦截器。

```
package com.bailiban.interceptor;

import org.springframework.web.servlet.HandlerInterceptor;

import javax.servlet.http.HttpServletRequest;
import javax.servlet.http.HttpServletResponse;
import javax.servlet.http.HttpSession;

/**
```

```
 * @author zhulang
 * @Classname LoginInterceptor
 * @Description 登录拦截器
 * @Date 2020/8/28 16:53
 */
public class LoginInterceptor implements HandlerInterceptor {
    @Override
    public boolean preHandle(HttpServletRequest request, HttpServletResponse response, Object handler)
throws Exception {
        HttpSession session = request.getSession();
        if (session.getAttribute("user") != null) {
            return true;
        }
// 用户没有登录跳转到登录页面
        request.getRequestDispatcher("/WEB-INF/view/login.jsp").forward(request, response);
        return false;
    }

}
```

(4) 配置拦截器。

```
<mvc:interceptors>
    <mvc:interceptor>
        <mvc:mapping path="/**"/>
        <mvc:exclude-mapping path="/emp/login"/>
        <mvc:exclude-mapping path="/css/**"/>
        <mvc:exclude-mapping path="/js/**"/>
        <mvc:exclude-mapping path="/image/**"/>
        <bean id="loginInterceptor" class="com.bailiban.interceptor.LoginInterceptor"/>
    </mvc:interceptor>
</mvc:interceptors>
```

上面配置中，通过<mvc:exclude-mapping>排除无须进行拦截的文件，这样就完成了登录拦截。

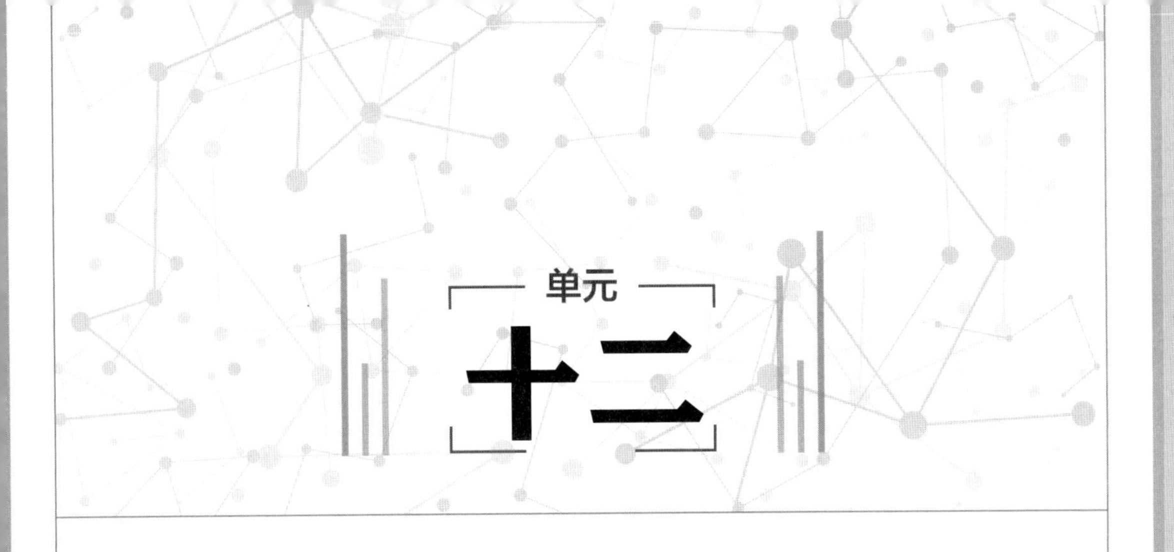


# Spring MVC项目开发

## 课程目标

- ❖ 了解项目介绍
- ❖ 了解业务介绍
- ❖ 掌握数据库的构建方法
- ❖ 掌握项目的构建方法
- ❖ 编写代码并测试

**简介**

通过对前边 11 个单元的学习，我们已经对 Spring、Spring MVC、MyBatis 有了深入的了解，从本单元开始将进入知识点的综合应用，将采用项目的形式把实际工作的内容复现于教学之中。

## 12.1 项目概述

“百里半门票销售系统”是一套售票系统，模拟完成一个 2022 年北京冬运会门票的销售系统，以提高大家的综合实战应用能力，本项目所展示的数据均是虚拟数据，主要包括注册、登录、个人信息管理、门票管理、我的订单、门票销售等功能模块。

### 12.1.1 项目特点

- 技术上采用 MyBatis、Spring、Spring MVC 来完成。
- 使用 rest-api 和 html 页面交互完成。
- 前后端数据交互，以标准的 JSON 格式数据完成。
- 使用主流的 MySql 数据库作为数据管理。

### 12.1.2 项目介绍

项目一共分为四个模块。

- 注册、登录模块，该模块完成新用户的注册，已注册用户登录等。
- 订单模块，该模块完成订单的新增、修改、删除、支付、历史信息查看等。
- 销售模块，包括下单、支付、取消等功能。
- 用户模块，包括修改用户信息、验证身份等。

### 12.1.3 环境要求

- 开发环境：JDK 1.8。
- 运行环境：Window Server 2003 以上、CentOS 7+(付费版 Linux 未测试)。
- 数据库：MySql 5.5+。
- 版本管理：Maven 3.6+。

### 12.1.4　数据交互

前端以 LayUI 和静态 HTML 为控制模板，使用 AJAX 与后台的 rest-api 以 JSON 数据为格式进行交互。以后台 Controller 为控制端，控制项目逻辑，如图 12-1 所示。

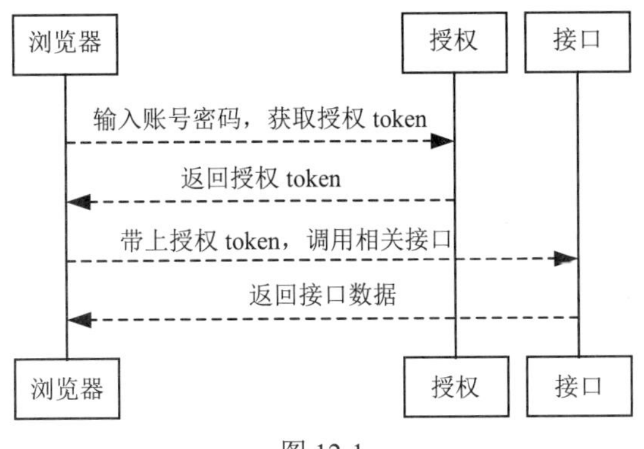


图 12-1

## 12.2 业务介绍

传统的软件开发，一般都是页面展示数据、后台处理数据、数据库持久化数据。换位思考，以一个业内人士的身份来看，所做的事情就是新增数据、删除数据、查询数据、修改数据，那么这些基本操作是否算是一个完整的项目？答案肯定是否定的。这些技术点如何结合才能算是一个完整的项目？这时就需要有对应的流程，来对增、删、查、改做出对应的梳理。而在一般的情景下对应的流程就是与客户的需求相匹配的流程。

### 12.2.1　业务流程

项目开发工程中最常见、同时又能够一目了然的就是业务流程图了，如图 12-2 所示为对“百里半门票销售系统”的业务梳理。

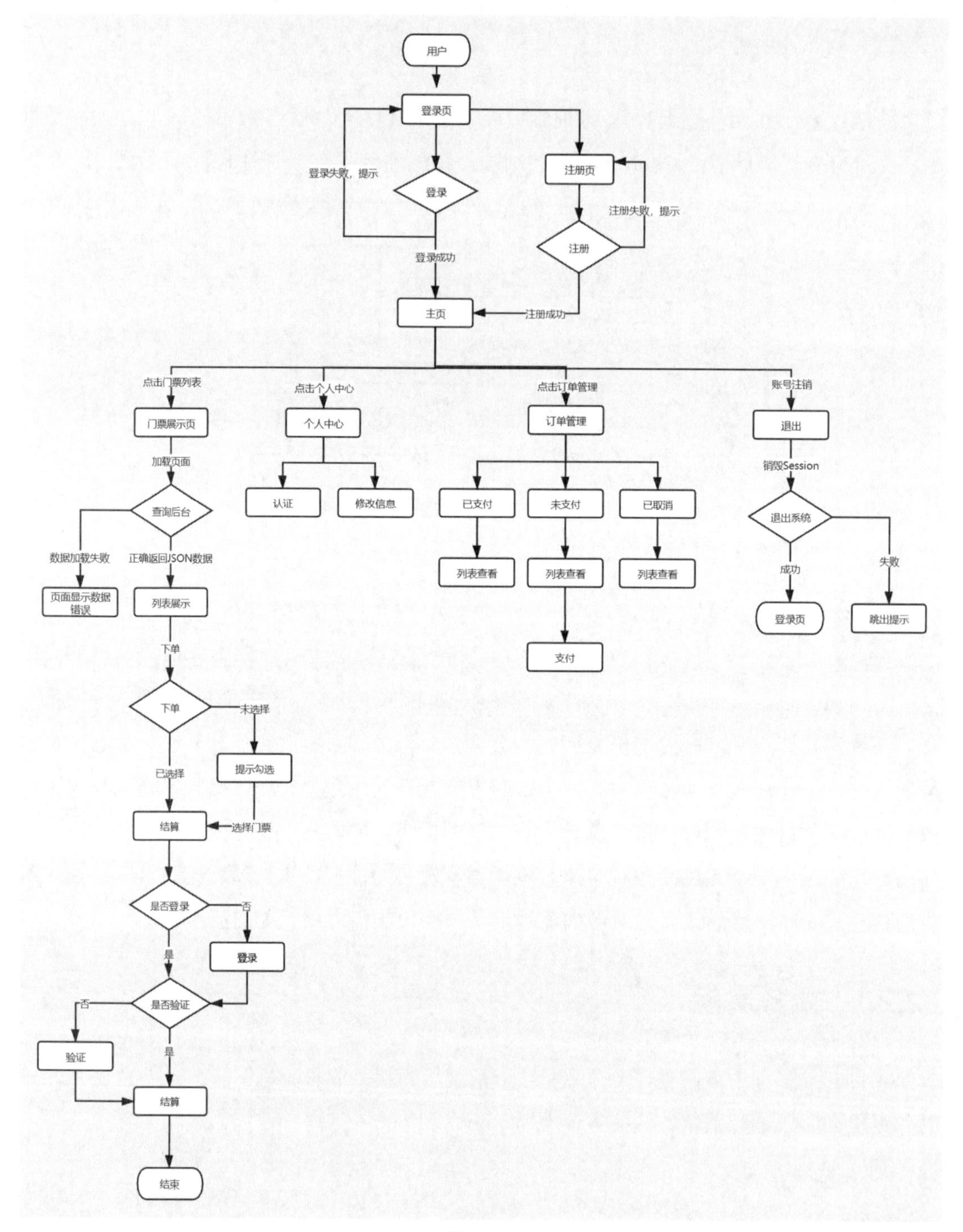


图 12-2

## 12.2.2 基础展示

用户登录页面如图 12-3 所示。

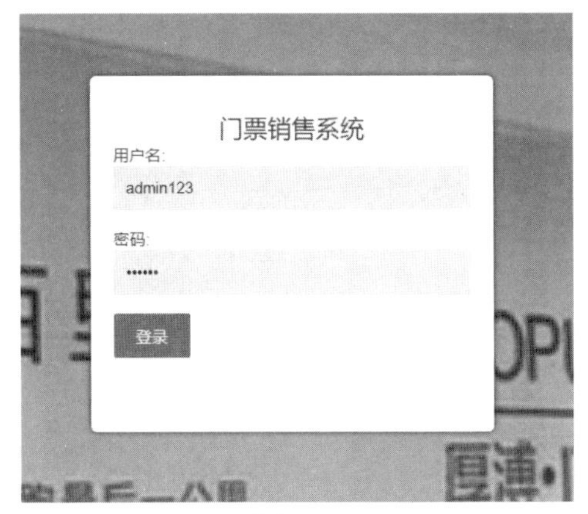


图 12-3

门票列表如图 12-4 所示。

首页　　我的订单　个人中心　登录　注册

| 场馆信息 | 编号 | 名称 | 开赛日期 | 余票数量 | 票价 | 状态 | 简介 |
|---|---|---|---|---|---|---|---|
| › | 1 | 第24届北京冬奥会开幕式门票 | 1643904000000 | 4999 | 2000 | 在售 | 令人期待的开幕式演出，赶紧来！ |
| › | 2 | 冰壶比赛门票 | 1643990400000 | 1999 | 1000 | 在售 | 冰上的“国际象棋”，快来看看！！！ |
| › | 3 | 男子冰球比赛门票 | 1644076800000 | 5997 | 1500 | 在售 | 冰上曲棍球，喜欢看球的抓紧买票 |
| › | 4 | 速滑比赛门票 | 1644076800000 | 8997 | 1600 | 在售 | 速度与激情的视觉盛宴 |
| › | 5 | 女子冰球比赛门票 | 1644163200000 | 6000 | 1500 | 在售 | 冰上曲棍球，喜欢看球的抓紧买票 |
| › | 6 | 冬季两项比赛门票 | 1644422400000 | 5 | 800 | 在售 | 冰上射击比赛，刺激与挑战并存 |

共 6 条　10条/页　‹ 1 ›　前往 1 页

取消选择

图 12-4

购票界面如图 12-5 所示。

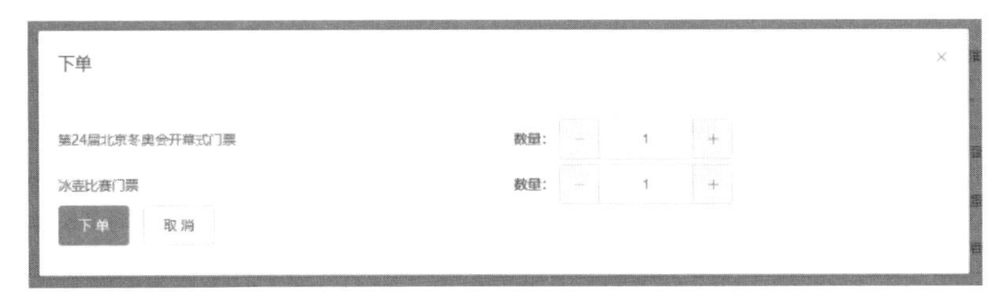


图 12-5

# 12.3 数据库设计

## 12.3.1 数据库介绍

根据需求以及业务流程可以分析出需要以下 5 张表：

- t_order_base(基础订单表)
- t_order_detail(订单详情表)
- t_stadium(场地表)
- t_ticket(门票表)
- t_user(用户表)

表之间的关系如图 12-6 所示。

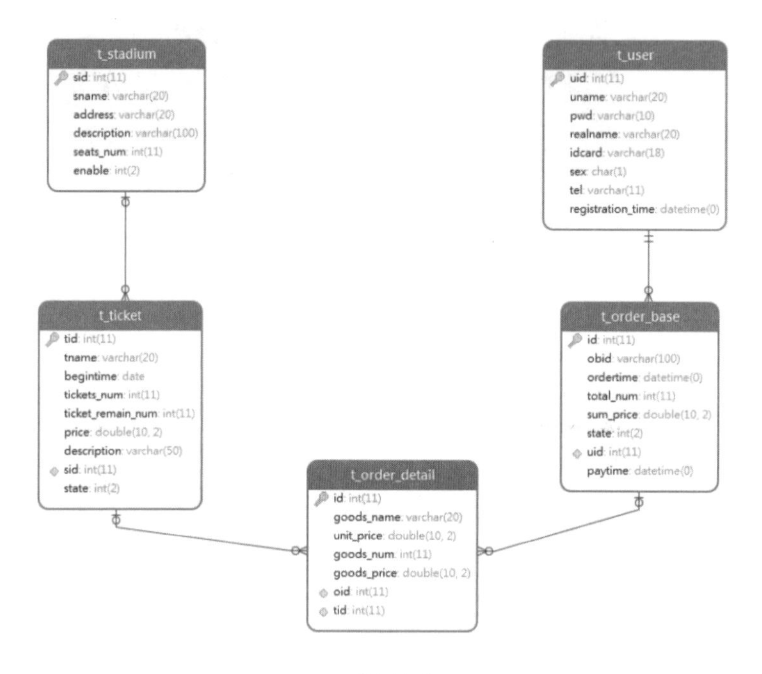


图 12-6

## 12.3.2 t_order_base(基础订单表)

基础订单表的代码如下。

```
CREATE TABLE `t_order_base` (
`id` INT(11) NOT NULL AUTO_INCREMENT,
`obid` VARCHAR(100) NOT NULL DEFAULT '' COMMENT '订单编号',
`ordertime` DATETIME DEFAULT NULL,
`total_num` INT(11) DEFAULT NULL COMMENT '购买总数量',
`sum_price` DOUBLE(10,2) DEFAULT NULL,
`state` INT(2) DEFAULT '0' COMMENT '0:待支付；1：已支付；-1：已删除',
`uid` INT(11) NOT NULL,
`paytime` DATETIME DEFAULT NULL ON UPDATE CURRENT_TIMESTAMP,
PRIMARY KEY (`id`),
KEY `o_u_fk` (`uid`),
CONSTRAINT `o_u_fk` FOREIGN KEY (`uid`) REFERENCES `t_user` (`uid`)
) ENGINE=INNODB AUTO_INCREMENT=21 DEFAULT CHARSET=utf8mb4;
```

```
INSERT INTO `t_order_base` VALUES ('17', 'f22f8723bcc846c0bd7a74da1ad4fcd9', '2020-09-15
        16:20:15', '4', '5500.00', '1', '5', '2020-09-15 16:20:18');
INSERT INTO `t_order_base` VALUES ('18', 'f820154d65a34236ad4635267a081c14', '2020-09-15
        16:23:00', '5', '4000.00', '-1', '5', '2020-09-15 16:28:28');
INSERT INTO `t_order_base` VALUES ('20', '513094466ee741c9903598a248e7401f', '2020-09-15
        16:40:04', '5', '4000.00', '-1', '5', '2020-09-15 16:40:56');
```

## 12.3.3 t_order_detail(订单详情表)

订单详情表的代码如下。

```
CREATE TABLE `t_order_detail` (
 `id` INT(11) NOT NULL AUTO_INCREMENT,
 `goods_name` VARCHAR(20) DEFAULT NULL,
 `unit_price` DOUBLE(10,2) DEFAULT NULL,
 `goods_num` INT(11) DEFAULT NULL,
 `goods_price` DOUBLE(10,2) DEFAULT NULL,
 `oid` INT(11) DEFAULT NULL,
 `tid` INT(11) DEFAULT NULL,
 PRIMARY KEY (`id`),
 KEY `od_ob_fk` (`oid`),
 KEY `od_t_fk` (`tid`),
 CONSTRAINT `od_ob_fk` FOREIGN KEY (`oid`) REFERENCES `t_order_base` (`id`),
 CONSTRAINT `od_t_fk` FOREIGN KEY (`tid`) REFERENCES `t_ticket` (`tid`)
) ENGINE=INNODB AUTO_INCREMENT=46 DEFAULT CHARSET=utf8mb4;

INSERT INTO `t_order_detail` VALUES ('41', '冰壶比赛门票', '1000.00', '1', '1000.00', '17', '2');
INSERT INTO `t_order_detail` VALUES ('42', '男子冰球比赛门票', '1500.00', '3', '4500.00', '17', '3');
INSERT INTO `t_order_detail` VALUES ('43', '冬季两项比赛门票', '800.00', '5', '4000.00', '18', '6');
INSERT INTO `t_order_detail` VALUES ('45', '冬季两项比赛门票', '800.00', '5', '4000.00', '20', '6');
```

## 12.3.4 t_stadium(场地表)

场地表的代码如下。

```
CREATE TABLE `t_stadium` (
 `sid` INT(11) NOT NULL AUTO_INCREMENT,
 `sname` VARCHAR(20) DEFAULT NULL,
 `address` VARCHAR(20) DEFAULT NULL,
 `description` VARCHAR(100) DEFAULT NULL,
 `seats_num` INT(11) DEFAULT NULL,
 `enable` INT(2) DEFAULT '1' COMMENT '0：不可用；1：可用',
 PRIMARY KEY (`sid`)
) ENGINE=INNODB AUTO_INCREMENT=13 DEFAULT CHARSET=utf8mb4;
```

```
INSERT INTO `t_stadium` VALUES ('1', '水立方', '北京奥林匹克公园内', '北京 2008 年夏季奥运会修
                        建的主游泳馆，将承担北京 2022 年冬奥会冰壶项目', '4500', '1');
INSERT INTO `t_stadium` VALUES ('2', '国家体育馆', '北京奥林匹克公园中心区的南部', '奥林匹克中
                        心区的标志性建筑之一,北京奥运会三大主场馆之一,将承担北
                        京 2022 年冬奥会男子冰球项目', '18000', '1');
INSERT INTO `t_stadium` VALUES ('3', '国家速滑馆', '北京奥林匹克公园西侧', '原是 2008 年北京奥
                        运会射箭场和曲棍球场的临时场馆，将承担北京 2022 年冬奥会
                        速度滑冰项目', '12000', '1');
INSERT INTO `t_stadium` VALUES ('4', '五棵松体育中心', '北京长安街和西四环路的交汇处', '北京奥
                        运会重要赛事场馆，将承担北京 2022 年冬奥会女子冰球项目',
                        '9000', '1');
', '将承担北京 2022 年冬奥会自由式滑雪及单板滑雪空中技巧、雪上技巧、U 型场地、坡面障碍技巧、
平行大回转、障碍追逐 6 个项目的比赛', '6500', '1');
```

## 12.3.5 t_ticket(门票表)

门票表的代码如下。

```
CREATE TABLE `t_ticket` (
 `tid` INT(11) NOT NULL AUTO_INCREMENT,
 `tname` VARCHAR(20) DEFAULT NULL,
 `begintime` DATE DEFAULT NULL COMMENT '比赛日期',
 `tickets_num` INT(11) DEFAULT NULL,
 `ticket_remain_num` INT(11) DEFAULT NULL,
 `price` DOUBLE(10,2) DEFAULT NULL,
 `description` VARCHAR(50) DEFAULT NULL,
 `sid` INT(11) DEFAULT NULL,
 `state` INT(2) DEFAULT '1' COMMENT '0：停售；1：在售',
 PRIMARY KEY (`tid`),
 KEY `t_s_fk` (`sid`),
 CONSTRAINT `t_s_fk` FOREIGN KEY (`sid`) REFERENCES `t_stadium` (`sid`)
) ENGINE=INNODB AUTO_INCREMENT=7 DEFAULT CHARSET=utf8mb4;

INSERT INTO `t_ticket` VALUES ('1', '第 24 届北京冬奥会开幕式门票', '2022-02-04', '8000', '5000',
                        '2000.00', '令人期待的开幕式演出，赶紧来！', '6', '1');
INSERT INTO `t_ticket` VALUES ('2', '冰壶比赛门票', '2022-02-05', '3000', '1999', '1000.00', '冰上的
                        “国际象棋”，快来看看！！！', '1', '1');
INSERT INTO `t_ticket` VALUES ('3', '男子冰球比赛门票', '2022-02-06', '12000', '5997', '1500.00', '冰上
                        曲棍球，喜欢看球的抓紧买票', '2', '1');
INSERT INTO `t_ticket` VALUES ('4', '速滑比赛门票', '2022-02-06', '9000', '8997', '1600.00', '速度与激
                        情的视觉盛宴', '3', '1');
```

## 12.3.6 t_user(用户表)

用户表的代码如下。

```
CREATE TABLE `t_user` (
 `uid` INT(11) NOT NULL AUTO_INCREMENT,
 `uname` VARCHAR(20) DEFAULT NULL,
 `pwd` VARCHAR(10) DEFAULT NULL,
 `realname` VARCHAR(20) DEFAULT NULL COMMENT '真实姓名',
 `idcard` VARCHAR(18) DEFAULT NULL,
 `sex` CHAR(1) DEFAULT NULL,
 `tel` VARCHAR(11) DEFAULT NULL,
 `registration_time` DATETIME DEFAULT NULL,
 PRIMARY KEY (`uid`)
) ENGINE=INNODB AUTO_INCREMENT=6 DEFAULT CHARSET=utf8mb4;

INSERT INTO `t_user` VALUES ('1', 'rose', 'rose123', '李柔丝', '420012003456321980', '女',
                                  '18966688800', '2020-09-09 16:26:22');
INSERT INTO `t_user` VALUES ('3', '浪哥', '123456', '朱浪', '421125199612234513', '男', '13397111234',
                                  '2020-09-15 15:53:06');
INSERT INTO `t_user` VALUES ('4', '百里半', '123456', '百里半', '421100199018093421', '男',
                                  '18966688800', '2020-09-13 16:57:08');
INSERT INTO `t_user` VALUES ('5', '厚溥', '123456', '厚溥教育', '421125199012123047', '男',
                                  '13388888800', '2020-09-15 16:07:05');
```

# 12.4 项目构建

可以采用 Maven 版本和 Jar 包版本两种方式来构建项目。

## 12.4.1　普通 Jar 包项目

### 1. 新建

右键单击 Eclipse，选择 New→Other 选项，如图 12-7 所示。

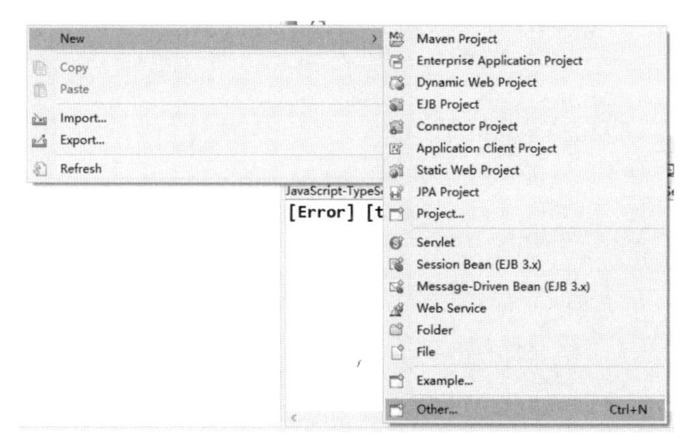


图 12-7

## 2. 选择项目

选择一个动态的 Web 项目，如图 12-8 所示。

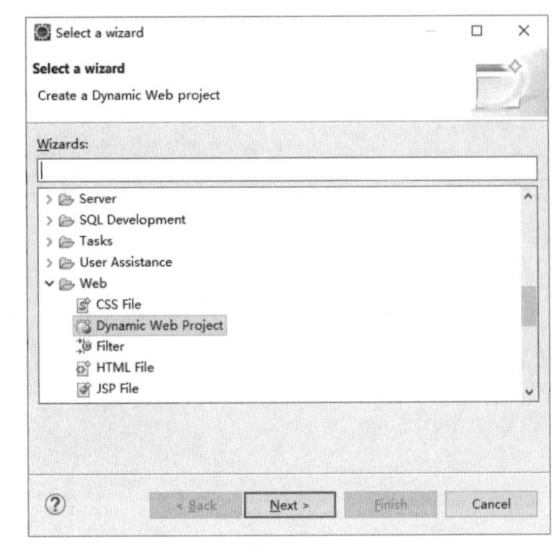


图 12-8

## 3. 完成构建

输入分组名字、项目名字，选择生成的内容，然后单击 Finish 按钮即可完成，如图 12-9 所示。

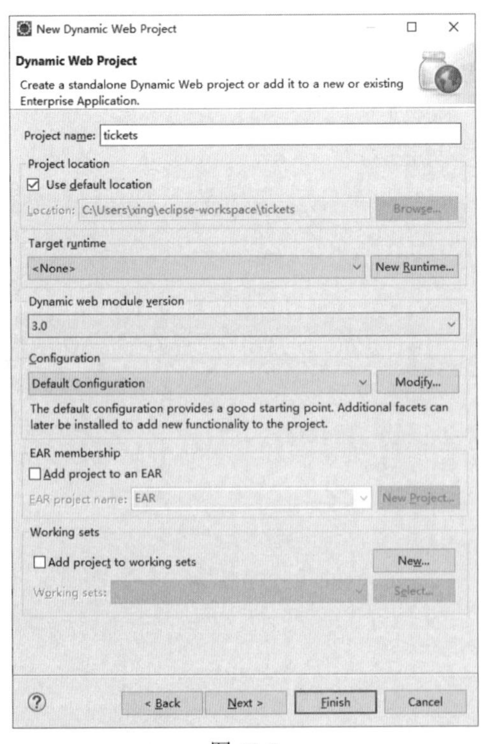


图 12-9

### 4. Jar 包

导入已经准备好的 Jar 包(图 12-10)。

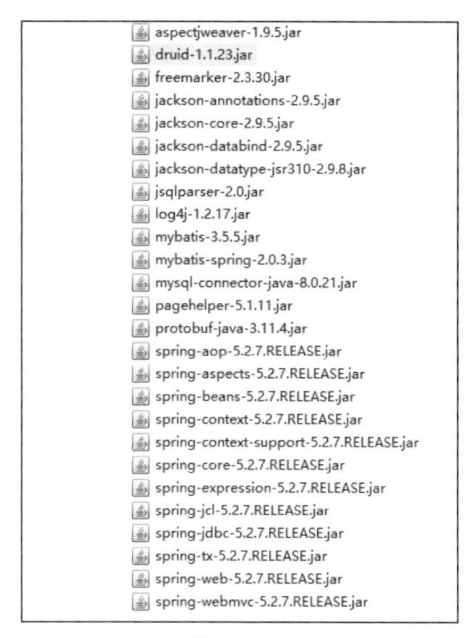


图 12-10

## 12.4.2　普通 Maven 项目

### 1. 新建

右键单击 Eclipse，选择 New→Other 选项(见图 12-7)，然后选择 Maven Project 选项，如图 12-11 所示。

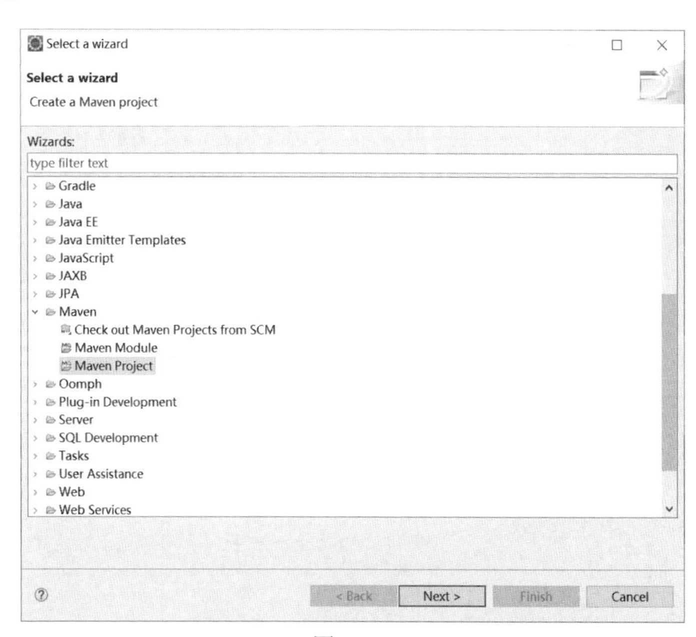


图 12-11

### 2. 选择内容

建立一个简单的 Maven 项目，如图 12-12 所示。

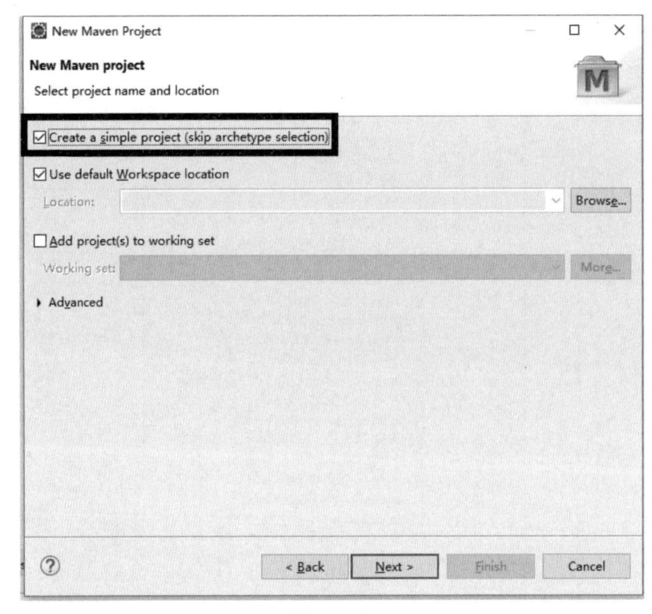


图 12-12

### 3. 完成

输入分组名字、项目名字，选择生成的内容，单击 Finish 按钮即可完成，如图 12-13 所示。

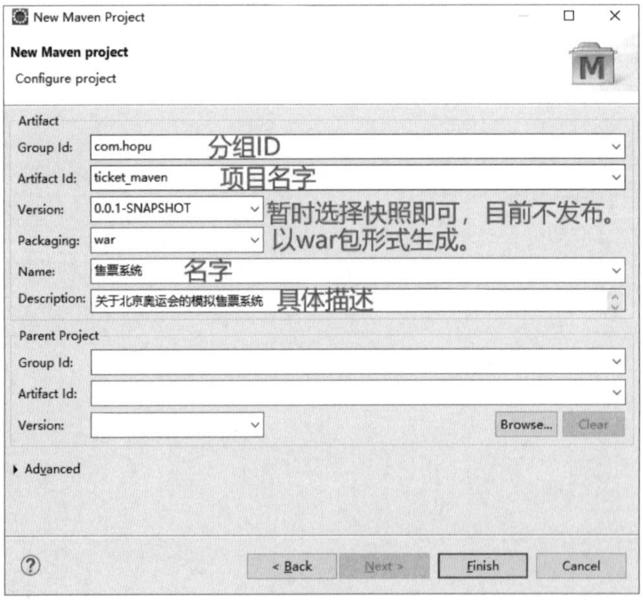

图 12-13

#### 4. Jar 包

把已经准备好的 pom 导入到项目内。

```
<groupId>com.hopu</groupId>
<artifactId>ticket_sales_system</artifactId>
<version>1.0-SNAPSHOT</version>
<packaging>war</packaging>
<name>ticket_sales_system Maven Webapp</name>
<!-- FIXME change it to the project's website -->
<url>http://www.example.com</url>
<properties>
    <project.build.sourceEncoding>UTF-8</project.build.sourceEncoding>
    <maven.compiler.source>1.8</maven.compiler.source>
    <maven.compiler.target>1.8</maven.compiler.target>
</properties>
<dependencies>
    <!--Spring 测试包-->
    <dependency>
        <groupId>org.springframework</groupId>
        <artifactId>spring-test</artifactId>
        <version>5.2.7.RELEASE</version>
        <scope>test</scope>
    </dependency>
    <!--测试包-->
    <dependency>
        <groupId>junit</groupId>
        <artifactId>junit</artifactId>
        <version>4.13</version>
        <scope>test</scope>
    </dependency>
    <!--springweb 依赖包-->
    <dependency>
        <groupId>org.springframework</groupId>
        <artifactId>spring-webmvc</artifactId>
        <version>5.2.7.RELEASE</version>
    </dependency>
    <!--jdbc 驱动包-->
    <dependency>
        <groupId>org.springframework</groupId>
        <artifactId>spring-jdbc</artifactId>
        <version>5.2.7.RELEASE</version>
    </dependency>
    <!--Spring 切面包-->
    <dependency>
        <groupId>org.springframework</groupId>
        <artifactId>spring-aspects</artifactId>
```

```
        <version>5.2.7.RELEASE</version>
    </dependency>
    <!--MyBatis 包-->
    <dependency>
        <groupId>org.mybatis</groupId>
        <artifactId>mybatis</artifactId>
        <version>3.5.5</version>
    </dependency>
    <!--MyBatis 和 Spring 整合包-->
    <dependency>
        <groupId>org.mybatis</groupId>
        <artifactId>MyBatis-Spring</artifactId>
        <version>2.0.3</version>
    </dependency>
    <!--分页插件依赖-->
    <dependency>
        <groupId>com.github.pagehelper</groupId>
        <artifactId>pagehelper</artifactId>
        <version>5.1.11</version>
    </dependency>
    <!--Servlet 的支持-->
    <dependency>
        <groupId>javax.servlet</groupId>
        <artifactId>servlet-api</artifactId>
        <version>2.5</version>
        <scope>provided</scope>
    </dependency>
    <!--连接池-->
    <dependency>
        <groupId>com.alibaba</groupId>
        <artifactId>druid</artifactId>
        <version>1.1.23</version>
    </dependency>
    <!--MySql 连接驱动-->
    <dependency>
        <groupId>mysql</groupId>
        <artifactId>mysql-connector-java</artifactId>
        <version>8.0.21</version>
    </dependency>
    <!--简写 get/set-->
    <dependency>
        <groupId>org.projectlombok</groupId>
        <artifactId>lombok</artifactId>
        <version>1.18.12</version>
        <scope>provided</scope>
    </dependency>
    <!--日志-->
```

```
    <dependency>
        <groupId>log4j</groupId>
        <artifactId>log4j</artifactId>
        <version>1.2.17</version>
    </dependency>
    <!--JSON 依赖-->
    <dependency>
        <groupId>com.fasterxml.jackson.core</groupId>
        <artifactId>jackson-databind</artifactId>
        <version>2.9.5</version>
    </dependency>
    <!--JSON 依赖-->
    <dependency>
        <groupId>com.fasterxml.jackson.core</groupId>
        <artifactId>jackson-annotations</artifactId>
        <version>2.9.5</version>
    </dependency>
    <!--freemarker，html 视图-->
    <dependency>
        <groupId>org.freemarker</groupId>
        <artifactId>freemarker</artifactId>
        <version>2.3.30</version>
    </dependency>
    <!--freemarker 依赖-->
    <dependency>
        <groupId>org.springframework</groupId>
        <artifactId>spring-context-support</artifactId>
        <version>5.2.7.RELEASE</version>
    </dependency>
    <!--日期格式化-->
    <dependency>
        <groupId>com.fasterxml.jackson.datatype</groupId>
        <artifactId>jackson-datatype-jsr310</artifactId>
        <version>2.9.8</version>
    </dependency>
</dependencies>
<build>
    <plugins>
        <plugin>
            <groupId>org.apache.tomcat.maven</groupId>
            <artifactId>tomcat8-maven-plugin</artifactId>
            <version>3.0-r1756463</version>
            <configuration>
                <server>tomcat8</server>
                <charset>utf-8</charset>
                <port>8080</port>
                <path>/</path>
```

```
            </configuration>
        </plugin>
    </plugins>
</build>
```

## 12.4.3 配置项目

### 1. web.xml

配置项目的 web.xml 和 Spring 的配置文件。

```
!--配置 Spring IoC 配置文件路径-->
  <context-param>
    <param-name>contextConfigLocation</param-name>
    <param-value>classpath*:spring/applicationContext.xml</param-value>
  </context-param>
  <!--配置 ContextLoaderListener，用以初始化 Spring IoC 容器-->
  <listener>
    <listener-class>org.springframework.web.context.ContextLoaderListener</listener-class>
  </listener>
  <!--过滤器解决 post 乱码-->
  <filter>
    <filter-name>CharacterEncodingFilter</filter-name>
    <filter-class>org.springframework.web.filter.CharacterEncodingFilter</filter-class>
    <init-param>
      <param-name>encoding</param-name>
      <param-value>utf-8</param-value>
    </init-param>
  </filter>
  <filter-mapping>
    <filter-name>CharacterEncodingFilter</filter-name>
    <url-pattern>/*</url-pattern>
  </filter-mapping>
  <!--配置前端控制器-->
  <servlet>
    <!--名称-->
    <servlet-name>dispatcher</servlet-name>
    <!-- Servlet 类-->
    <servlet-class>org.springframework.web.servlet.DispatcherServlet</servlet-class>
    <init-param>
      <!-- Spring MVC 配置参数文件的位置-->
      <param-name>contextConfigLocation</param-name>
      <!-- (可不配置)默认名称为 ServletName-servlet.xml,即:dispatcher-servlet.xml-->
      <param-value>classpath*:spring/springmvc.xml</param-value>
    </init-param>
    <!--启动顺序，数字越小，启动越早，非 0 正整数-->
```

```
        <load-on-startup>1</load-on-startup>
    </servlet>
    <!-- Servlet 拦截配置-->
    <servlet-mapping>
        <servlet-name>dispatcher</servlet-name>
        <url-pattern>/</url-pattern>
    </servlet-mapping>
```

2. Spring 的配置

配置 Spring 的配置文件 applicationContext.xml，配置数据源、插件等。

```
<?xml version="1.0" encoding="UTF-8"?>
<beans xmlns="http://www.springframework.org/schema/beans"
       xmlns:xsi="http://www.w3.org/2001/XMLSchema-instance"
       xmlns:context="http://www.springframework.org/schema/context"
       xmlns:tx="http://www.springframework.org/schema/tx"
       xsi:schemaLocation="http://www.springframework.org/schema/beans
                           http://www.springframework.org/schema/beans/spring-beans.xsd
                           http://www.springframework.org/schema/context
                           http://www.springframework.org/schema/context/spring-context.xsd
                           http://www.springframework.org/schema/tx
                           http://www.springframework.org/schema/tx/spring-tx.xsd">
    <!--读取 properties 文件-->
    <context:property-placeholder location="classpath:properties/db.properties"/>
    <!--开启注解自动扫描-->
    <context:component-scan base-package="com.bailiban"/>
    <!--配置数据源-->
    <bean id="dataSource" class="com.alibaba.druid.pool.DruidDataSource">
        <property name="driverClassName" value="${jdbc.driver}"/>
        <property name="url" value="${jdbc.url}"/>
        <property name="username" value="${jdbc.username}"/>
        <property name="password" value="${jdbc.password}"/>
        <!-- 配置初始化大小、最小、最大 -->
        <property name="initialSize" value="2"/>
        <property name="minIdle" value="1"/>
        <property name="maxActive" value="10"/>
        <!-- 配置获取连接等待超时的时间 -->
        <property name="maxWait" value="60000"/>
    </bean>
    <!--MyBatis 配置-->
    <bean id="sqlSessionFactory" class="org.mybatis.spring.SqlSessionFactoryBean">
        <property name="dataSource" ref="dataSource"/>
        <property name="configLocation" value="classpath:mybatis/mybatis-config.xml"/>
        <!--MyBatis 插件-->
        <property name="plugins">
            <set>
                <!--配置 pageHelper 分页插件-->
```

```
                    <bean class="com.github.pagehelper.PageInterceptor">
                        <property name="properties">
                            <props>
                                <!--方言：-->
                                <prop key="helperDialect">mysql</prop>
                                <prop key="reansonable">true</prop>
                            </props>
                        </property>
                    </bean>
                </set>
            </property>
        </bean>
        <bean class="org.mybatis.spring.mapper.MapperScannerConfigurer">
            <property name="basePackage" value="com.hopu.mapper"/>
            <property name="sqlSessionFactoryBeanName" value="sqlSessionFactory"/>
        </bean>
        <!--事务管理器配置数据源事务-->
        <bean id="transactionManager"
         class="org.springframework.jdbc.datasource.DataSourceTransactionManager">
            <property name="dataSource" ref="dataSource"/>
        </bean>
        <!--使用注解定义事务-->
        <tx:annotation-driven/>
</beans>
```

### 3. 配置数据库的参数文件 db.properties

代码如下。

```
jdbc.driver=com.mysql.cj.jdbc.Driver
jdbc.url=jdbc:mysql://localhost:3306/ticket?serverTimezone=GMT%2b8&useSSL=false&
        useUnicode=true&characterEncoding=utf-8
jdbc.username=root
jdbc.password=123456
```

### 4. Spring MVC 的配置

springmvc.xml 配置静态文件、前缀后缀、拦截器等。

```
<?xml version="1.0" encoding="UTF-8"?>
<beans xmlns="http://www.springframework.org/schema/beans"
        xmlns:mvc="http://www.springframework.org/schema/mvc"
        xmlns:xsi="http://www.w3.org/2001/XMLSchema-instance"
        xmlns:context="http://www.springframework.org/schema/context"
        xsi:schemaLocation="
         http://www.springframework.org/schema/beans
         http://www.springframework.org/schema/beans/spring-beans.xsd
         http://www.springframework.org/schema/context
```

```
            http://www.springframework.org/schema/context/spring-context.xsd
            http://www.springframework.org/schema/mvc
            http://www.springframework.org/schema/mvc/spring-mvc.xsd">
    <!--定义扫描装载的包-->
    <context:component-scan base-package="com.bailiban.controller"/>
</beans>
```

5. mybatis-config.xml

配置对应的 MyBatis 的配置文件。

```
<!DOCTYPE configuration PUBLIC "-//mybatis.org//DTD Config 3.0//EN"
 "http://mybatis.org/dtd/mybatis-3-config.dtd">
<configuration><!--配置-->
    <mappers>
    </mappers>
</configuration>
```

# 12.5 编写代码并测试

## 12.5.1 编写 Controller

编写一个 Controller 测试，项目建立成功。写一个 API，用来做测试。

```
@Controller
@RequestMapping("user")
public class UserController {

/**
 * @auth: hopu
 * @return：String
 * @description：测试
 */
@RequestMapping("/getTest")
@ResponseBody
public String getTest() {
    return "test";
}

}
```

## 12.5.2 加载项目

(1) 把项目添加到 Tomcat 中，如图 12-14 所示。

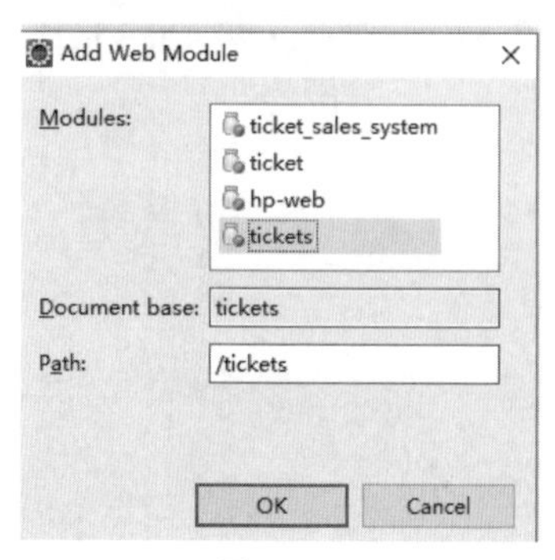


图 12-14

(2) 启动 Tomcat。

选择 Start 选项，即可启动 Tomcat，如图 12-15 所示。

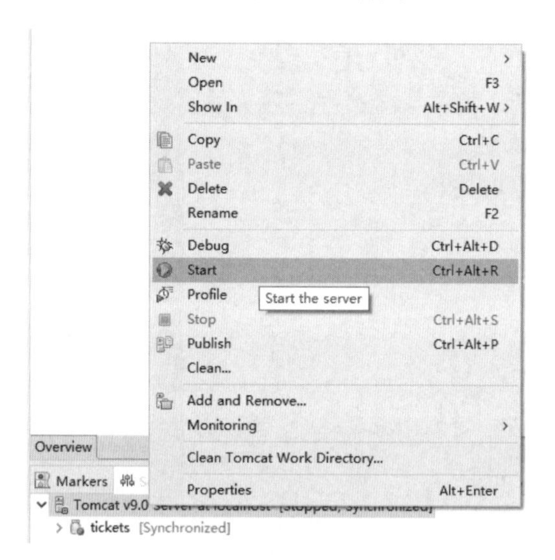


图 12-15

## 12.5.3 测试项目

使用测试工具 Postman 测试接口。获取如图 12-16 所示的结果，表明项目建立成功。

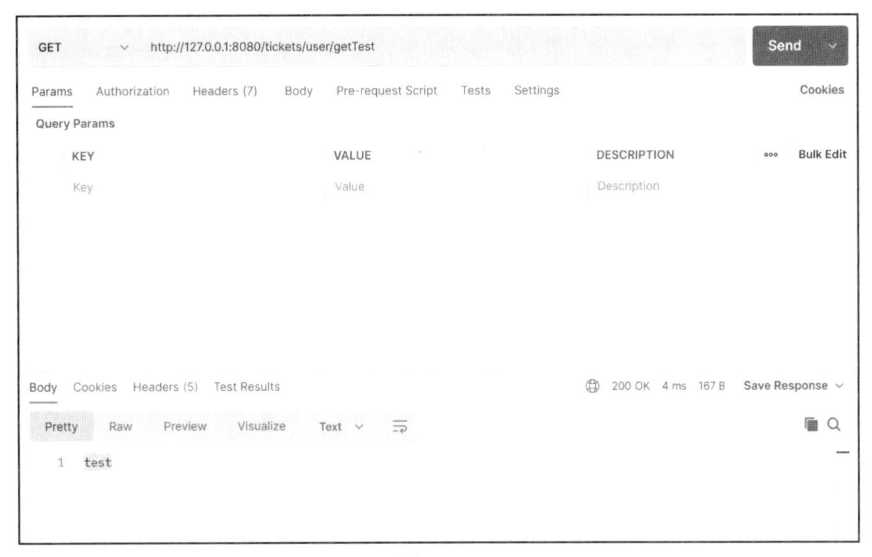


图 12-16

## 单元小结

- 了解“百里半门票销售系统”基本的项目特点，了解具体的项目内容，具备基础的业务分析能力，了解最基本的业务原型。
- 分析“百里半门票销售系统”的基本业务流程、数据库关系，构建基本的数据库，同时新增一部分基础数据用以测试。
- 构建 Spring、Spring MVC、MyBatis 结合的项目架构，并且能够启动成功，使用测试工作或者浏览器访问测试项目。

# 业务的抽象

## 课程目标

❖ 了解业务抽象

❖ 掌握信息配置方法

❖ 掌握持久层应用

❖ 掌握业务层应用

❖ 掌握控制层应用

**简介**

前一单元讲解了整个项目的基本概况，构建了项目的数据库，梳理了项目的基础业务逻辑，也同步完成了项目基础框架的构建、项目的 API 的基础测试。在本单元即将抽象化项目基础的业务逻辑，完成项目大部分 API 的设计以及编写。

# 13.1 抽象业务

“百里半门票销售系统”主要涉及业务、注册、登录、个人信息管理、门票管理、我的订单、门票销售等功能模块。比较老的 JSP 类的项目以 JSP 模板为页面输出，通过进行请求转发来做业务的控制。耦合性非常强是一体化项目的弊端，不利于项目的稳定性以及项目的健壮性。

## 13.1.1 业务功能图

业务功能所涉及的基础业务划分如图 13-1 所示。

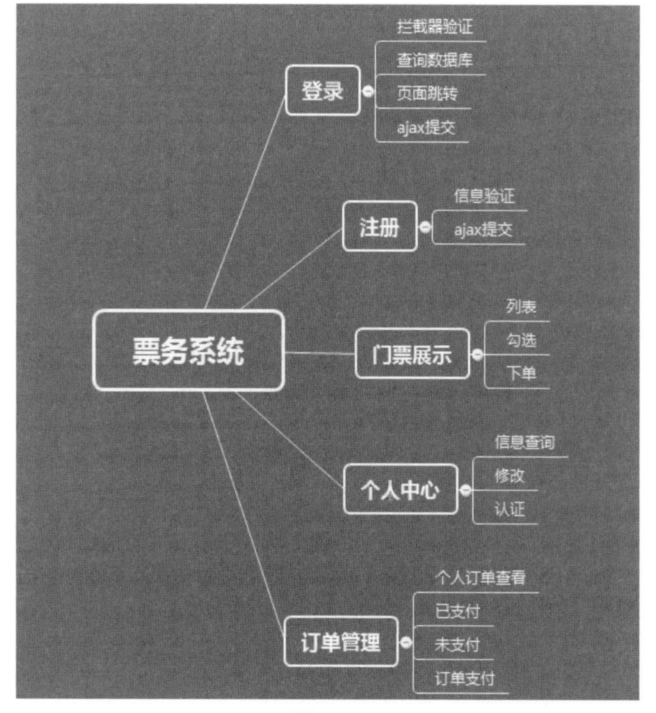


图 13-1

### 13.1.2 数据库关系图

根据需求以及业务流程可以分析出该项目需要以下 5 个表。

- t_order_base(基础订单表)
- t_order_detail(订单详情表)
- t_stadium(场地表)
- t_ticket(门票表)
- t_user(用户表)

表之间的关系如图 13-2 所示。

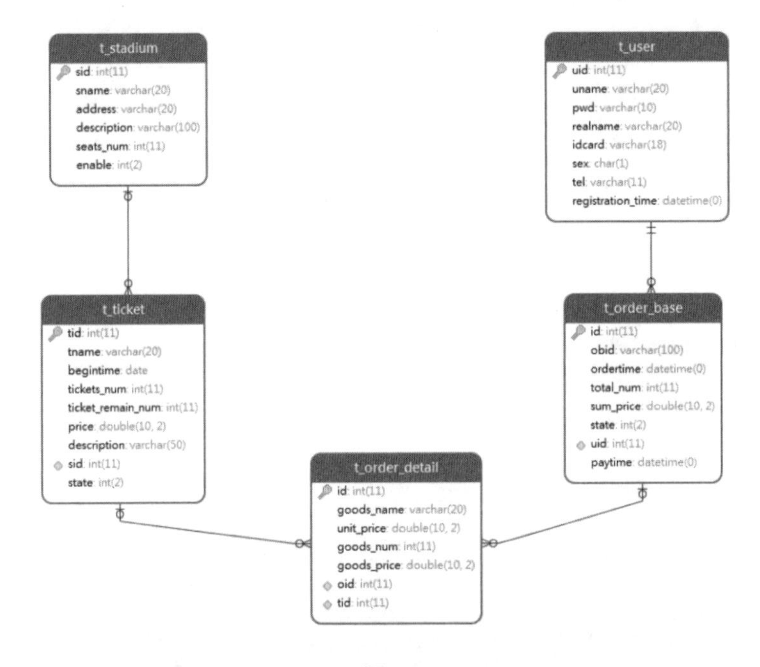


图 13-2

## 13.2 配置项目

紧接上一单元对项目进行进一步的推进。对 Spring 的配置文件做进一步处理，加入 MyBatis 的配置文件，增加 Spring MVC 的配置文件。

### 13.2.1 Spring 的配置修改

修改 Spring 的配置文件，applicationContext.xml 中的代码如下：

```
<!--配置 MyBatis -->
<bean id="sqlSessionFactory" class="org.mybatis.spring.SqlSessionFactoryBean">
    <property name="dataSource" ref="dataSource"/>
    <property name="configLocation" value="classpath:config/mybatis-config.xml"/>
     <!--MyBatis 插件-->
    <property name="plugins">
        <set>
            <!--配置 pageHelper 分页插件-->
            <bean class="com.github.pagehelper.PageInterceptor">
                <property name="properties">
                    <props>
                        <!--方言：-->
                        <prop kcy="helperDialect">mysql</prop>
                        <prop key="reansonable">true</prop>
                    </props>
                </property>
            </bean>
        </set>
    </property>
</bean>
<bean class="org.mybatis.spring.mapper.MapperScannerConfigurer">
    <property name="basePackage" value="com.hopu.mapper"/>
    <property name="sqlSessionFactoryBeanName" value="sqlSessionFactory"/>
</bean>
<!--事务管理器配置数据源事务-->
<bean id="transactionManager"
class="org.springframework.jdbc.datasource.DataSourceTransactionManager">
    <property name="dataSource" ref="dataSource"/>
</bean>
```

## 13.2.2 新增 MyBatis 的配置

补充 MyBatis 的配置文件，内容如下：

```
<configuration><!--配置-->
    <settings>
        <!--指定 MyBatis 使用 log4j -->
        <setting name="logImpl" value="LOG4J"/>
        <!--开启延迟加载功能-->
        <setting name="lazyLoadingEnabled" value="true"/>
        <!--不进行完整对象的加载，属性按需加载-->
        <setting name="aggressiveLazyLoading" value="false"/>
    </settings>
    <typeAliases>
        <!--类型别名-->
        <package name="com.bailiban.entity"/>
```

```
    </typeAliases>
    <mappers>
        <mapper resource="com/hopu/mapper/mappers/OrderBaseMapper.xml"/>
        <mapper resource="com/hopu/mapper/mappers/OrderDetailMapper.xml"/>
        <mapper resource="com/hopu/mapper/mappers/StadiumMapper.xml"/>
        <mapper resource="com/hopu/mapper/mappers/TicketMapper.xml"/>
        <mapper resource="com/hopu/mapper/mappers/UserMapper.xml"/>
    </mappers>
</configuration>
```

### 13.2.3 补充 Spring MVC 的配置

补充 Spring MVC 的配置文件，内容如下：

```
<!--定义扫描装载的包-->
    <context:component-scan base-package="com.hopu"/>
    <!--使用注解驱动-->
    <mvc:annotation-driven>
        <mvc:message-converters>
            <bean class="org.springframework.http.converter.json.MappingJackson2HttpMessageConverter"/>
        </mvc:message-converters>
    </mvc:annotation-driven>
```

## 13.3 API

完成数据库设计的同时也补充了配置文件的信息，下一步需要把数据库的数据通过项目传递给展示的页面，在 Spring 框架的逻辑构成中需要完成，从 DataBase 开始至持久层、持久层至业务逻辑层、业务逻辑层至控制层的一整套过程。

在项目开发的过程中使用了简化实体类开发的 lombok 工具，在使用 lombok 工具之后可以直接省略 get、set 方法，但需要使用注解@Data。

### 13.3.1 实体类的构建

首先根据数据库构建对应的实体类 User。

```
@Data
public class User implements Serializable {
    private static final long serialVersionUID = 503929492645954280L;

    private Integer uid;
```

```
    private String uname;

    private String pwd;
    /**
     * 真实姓名
     */
    private String realname;

    private String idcard;

    private String sex;

    private String tel;
    @JsonFormat(shape = JsonFormat.Shape.STRING, pattern = "yyyy-MM-dd HH:mm:ss")
    @DateTimeFormat(pattern = "yyyy-MM-dd HH:mm:ss")
    private LocalDateTime registrationTime;
}
```

## 13.3.2　持久层的构建

构建用户模块的数据持久层，也就是常说的 Dao 层。首先思考如何构建用户需要的接口信息设计基础的项目标准，持久层信息如下。

```
@Mapper
@Repository
public interface UserMapper {

    /**
     * 通过 ID 查询单条数据
     *
     * @param uid  主键
     * @return  实例对象
     */
    User queryById(Integer uid);

    /**
     * 新增数据
     *
     * @param user  实例对象
     * @return  影响行数
     */
    int insert(User user);

    /**
     * 修改数据
```

```
     *
     * @param user 实例对象
     * @return 影响行数
     */
    int update(User user);

    /**
     * 根据 uname 和 pwd 查询用户
     *
     * @param uname 用户名
     * @param pwd    密码
     * @return user
     */
    User findUserByUnameAndPwd(@Param("uname") String uname, @Param("pwd") String pwd);
}
```

完成持久层后，需要完成与持久层接口对应的 Mapper 文件，具体的 Mapper 文件信息如下。

```
<mapper namespace="com.hopu.mapper.UserMapper">
    <resultMap type="com.hopu.entity.User" id="UserMap">
        <result property="uid" column="uid" jdbcType="INTEGER"/>
        <result property="uname" column="uname" jdbcType="VARCHAR"/>
        <result property="pwd" column="pwd" jdbcType="VARCHAR"/>
        <result property="realname" column="realname" jdbcType="VARCHAR"/>
        <result property="idcard" column="idcard" jdbcType="VARCHAR"/>
        <result property="sex" column="sex" jdbcType="VARCHAR"/>
        <result property="tel" column="tel" jdbcType="VARCHAR"/>
        <result property="registrationTime" column="registration_time" jdbcType="TIMESTAMP"/>
    </resultMap>
    <!--查询单个-->
    <select id="queryById" resultMap="UserMap">
        select uid,
   uname,
   pwd,
   realname,
   idcard,
   sex,
   tel,
   registration_time
        from t_user
        where uid = #{uid}
    </select>

    <!--新增所有列-->
    <insert id="insert" keyProperty="uid" useGeneratedKeys="true">
        insert into t_user(uname, pwd, realname, idcard, sex, tel, registration_time)
```

```
        values (#{uname}, #{pwd}, #{realname}, #{idcard}, #{sex}, #{tel}, #{registrationTime})
    </insert>

    <!--通过主键修改数据-->
    <update id="update">
        update t_user
        <set>
            <if test="uname != null and uname != ''">
            uname = #{uname},
            </if>
            <if test="pwd != null and pwd != ''">
            pwd = #{pwd},
            </if>
            <if test="realname != null and realname != ''">
            realname = #{realname},
            </if>
            <if test="idcard != null and idcard != ''">
            idcard = #{idcard},
            </if>
            <if test="sex != null and sex != ''">
            sex = #{sex},
            </if>
            <if test="tel != null and tel != ''">
            tel = #{tel},
            </if>
            <if test="registrationTime != null">
            registration_time = #{registrationTime},
            </if>
        </set>
        where uid = #{uid}
    </update>

    <!--根据用户名和密码查询用户-->
    <select id="findUserByUnameAndPwd" resultType="User">
        select uid,
   uname,
   pwd,
   realname,
   idcard,
   sex,
   tel,
   registration_time
        from t_user
        where uname = #{uname}
          and pwd = #{pwd}
    </select>
</mapper>
```

## 13.3.3 业务逻辑层的构建

用户模块持久层查询了数据库，完成了与数据库的交互，业务层需要调用持久层完成对数据的传递。

接口信息如下：

```
public interface UserService {
    /**
     * 通过 ID 查询单条数据
     *
     * @param uid 主键
     * @return 实例对象
     */
    User queryById(Integer uid);

    /**
     * 新增数据
     *
     * @param user 实例对象
     * @return 实例对象
     */
    User insert(User user);

    /**
     * 修改数据
     *
     * @param user 实例对象
     * @return 实例对象
     */
    User update(User user);

    /**
     * 根据用户名和密码查找用户
     *
     * @param uname 用户名
     * @param pwd     密码
     * @return
     */
    User findUserByUnameAndPwd(String uname, String pwd);
}
```

有了接口必然有与之对应的实现类。

```
@Service("userService")
@Transactional(rollbackFor = Exception.class)
public class UserServiceImpl implements UserService {
```

```
@Resource
private UserMapper userMapper;

/**
 * 通过 ID 查询单条数据
 *
 * @param uid  主键
 * @return  实例对象
 */
@Override
public User queryById(Integer uid) {
    return this.userMapper.queryById(uid);
}

/**
 * 新增数据
 *
 * @param user  实例对象
 * @return  实例对象
 */
@Override
public User insert(User user) {
    this.userMapper.insert(user);
    return user;
}

/**
 * 修改数据
 *
 * @param user  实例对象
 * @return  实例对象
 */
@Override
public User update(User user) {
    this.userMapper.update(user);
    return this.queryById(user.getUid());
}

/**
 * 根据用户名和密码查找用户
 *
 * @param uname  用户名
 * @param pwd      密码
 * @return
 */
@Override
public User findUserByUnameAndPwd(String uname, String pwd) {
```

```
            return userMapper.findUserByUnameAndPwd(uname, pwd);
        }
    }
```

### 13.3.4 控制层的构建

完成用户业务层的编写后，需要把数据传递给控制层，由控制层最后通过接口把数据传递给展示层。

控制层的编写过程如下：

```
    @RestController
    @RequestMapping("user")
    public class UserController {
        /**

       * 服务对象
         /
            @Resource
            private UserService userService;

        /**
         * 登录提交
         *
         * @param user user
         * @return
         */
        @PostMapping("login")
        public Result login(HttpSession session, @RequestBody(required = false) User user) {

// 输出用户名和密码
            System.out.println("uname:" + user.getUname() + "---pwd:" + user.getPwd());
// 非空校验
            if (user == null || StringUtils.isEmpty(user.getUname()) || StringUtils.isEmpty(user.getPwd())) {
                System.out.println("登录失败");
                return Result.failure(ResultCode.PARAM_IS_BLANK);
            } else {
// 查询用户名和密码是否匹配
                user = userService.findUserByUnameAndPwd(user.getUname(), user.getPwd());
                if (user != null) {
// 匹配成功
        session.setAttribute("user", user);
        System.out.println("登录成功");
        return Result.success();
                } else {
// 匹配失败
```

```
        System.out.println("登录失败");
        return Result.failure(ResultCode.USER_LOGIN_ERROR);
                }
            }

        }

        /**
         * 用户注册
         *
         * @param user user
         * @return
         */
        @PostMapping("regist")
        public Result regist(HttpSession session, @RequestBody(required = false) User user) {

//  非空校验
            if (user != null) {
//  赋予注册时间
                user.setRegistrationTime(LocalDateTime.now());
                userService.insert(user);
//  注册成功，存入 session 标记
                session.setAttribute("registOK", user);
                return Result.success();
            }
            return Result.failure(ResultCode.PARAM_IS_BLANK);
        }


        /**
         * 用户校验、更新证件，也就是完善用户的身份信息
         *
         * @param user user
         * @return
         */
        @PostMapping("bindCard")
        public Result bindCard(HttpSession session, @RequestBody(required = false) User user) {

//  非空校验，当前 user 中只有真实姓名和身份证号
            if (user != null) {
//  更新证件信息，可以在注册之后立刻校验绑定身份信息，也可以选择登录后再绑定
                User registOK = (User) session.getAttribute("registOK");
                User loginUser = (User) session.getAttribute("user");
                if (registOK != null) {
//  如果是注册后绑定，此时处于未登录状态，从注册信息中获取 uid
        user.setUid(registOK.getUid());
                } else {
```

```
// 否则从登录信息中获取 id
    user.setUid(loginUser.getUid());
            }
// 更新以后的信息，也就是绑定身份信息
            user = userService.update(user);
// 绑定后重新更新 session 中的 user
            session.setAttribute("user", user);
            return Result.success();
        }
        return Result.failure(ResultCode.PARAM_IS_BLANK);
    }

    /**
     * 判断用户是否登录
     *
     * @param session session
     * @return
     */
    @GetMapping("isLogin")
    public Result isLogin(HttpSession session) {

// 从 session 中获取 user
        User user = (User) session.getAttribute("user");
        if (user != null) {

            if (StringUtils.isEmpty(user.getIdcard())) {

// 用户已注册，但是身份信息未校验
    return Result.failure(ResultCode.USER_NOT_CHECKED);
            } else {
    return Result.success();
            }
        } else {
            return Result.failure(ResultCode.USER_NOT_LOGGED_IN);
        }
    }

    /**
     * 用户个人信息
     *
     * @param session session
     * @return
     */
    @GetMapping("myinfo")
    public Result info(HttpSession session) {

// 直接从 session 中获取用户的 id
```

```
            User user = (User) session.getAttribute("user");
// 根据 id 查询用户
            user = userService.queryById(user.getUid());
            return Result.success(user);
        }
}
```

## 13.3.5　测试项目

选择 Start 选项，构建用户模块启动项目，如图 13-3 所示。

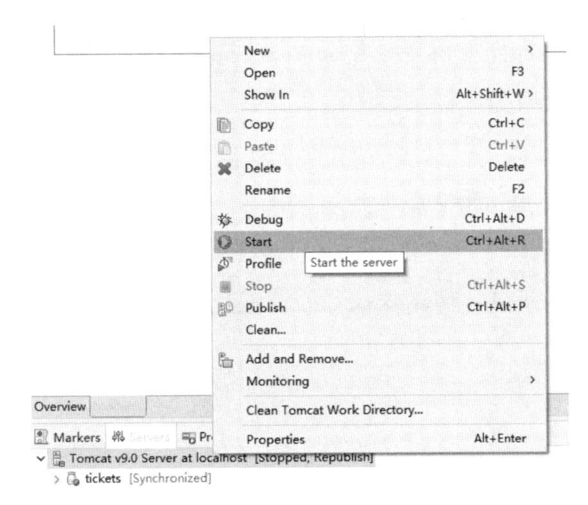


图 13-3

打开 API 的测试工具 Postman，输入项目的访问地址，如图 13-4 所示。

图 13-4

## 单元小结

- 本单元的内容比较重要，几乎包括了本次项目的所有业务，而且完成了前后端的控制反转的容器、项目的基础操作。
- 当下的项目基本都是前后端分离，或者多种设备共用一个 Server(服务器)，已经很少有十多年前流行的 B/S 模型直接使用后台直接输出 Browser(浏览器)的脚本。
- API 是前后分离后台最后的节点，本次项目并没有做完全的前后端分离，仅仅是做了伪分离，前后端分离往往控制权在后台，运行的逻辑都在页面。
- dao 层的 CURD 操作如何复合至 service 层，service 层如何执行业务复合处理，最后由 Controller 控制，做最后的输出，每一种操作对应的都是接口，不再是所见即所得。
- 每个业务之间无直接的关联，这就相当于业务的解耦，项目变成了一个容器，前台需要什么告诉后台，然后后台返回给前台所要的内容。至于后台如何完成，不是前台控制的，这就是所谓的业务的控制反转。

# 前后端交互

## 课程目标

- ❖ 掌握登录功能展示模块的设计方法
- ❖ 掌握页面控制类的应用
- ❖ 掌握项目的编写方法
- ❖ 掌握拦截器的应用

**简介**

在上一单元构建了项目的 API，完成了对接口的增、删、查、改等操作，项目中仅能在测试工具和页面看到对应的 JSON 数据，不能直观地观察到 View 的变化，但是项目给用户可操作的部分在日常中都是有界面的，可以直观地展示于各个终端。

# 14.1 业务功能

“百里半门票销售系统”所使用的 View 部分以静态 HTML、JavaScript、CSS 等作为主要技术。项目以教学为最终目标，使用了 MVC 的模式进行开发。HTML、JavaScript、CSS 静态文件继续存放于项目之内。

注册页面如图 14-1 所示。

图 14-1

登录页面如图 14-2 所示。

图 14-2

## 14.2 静态文件

JSP 页面是动态地从 JavaWeb 项目中直接输出脚本，而 HTML 文件是静态的文件。如果将静态文件直接放入使用 Spring MVC 框架的项目，同时又使用 Restful 风格的 url 拦截，在 web.xml 文件中直接配置“ / ”，这时所有的文件都会被 Spring MVC 框架拦截，项目内的所有静态文件都会处于不可被访问状态，静态文件被拦截可以使用三种方式来解决。

web.xml 的 Spring MVC 配置如下：

```
<!--配置前端控制器-->
  <servlet>
    <!--名称-->
    <servlet-name>dispatcher</servlet-name>
    <!-- Servlet 类-->
    <servlet-class>org.springframework.web.servlet.DispatcherServlet</servlet-class>
    <init-param>
      <!-- Spring MVC 配置参数文件的位置-->
      <param-name>contextConfigLocation</param-name>
      <!-- (可不配置)默认名称为 ServletName-servlet.xml，即 dispatcher-servlet.xml-->
      <param-value>classpath*:config/springmvc.xml</param-value>
    </init-param>
    <!--启动顺序，数字越小，启动越早，为非 0 正整数-->
    <load-on-startup>1</load-on-startup>
  </servlet>
```

```
<!-- Servlet 拦截配置-->
<servlet-mapping>
   <servlet-name>dispatcher</servlet-name>
   <url-pattern>/</url-pattern>
</servlet-mapping>
```

解决方案如下：

(1) 激活中间件的默认 Servlet 来处理静态文件。

(2) 使用在 Spring 3.0.4 以后版本中提供的 mvc：resources。

(3) 使用 mvc:default-servlet-handler/。

### 14.2.1 激活中间件的默认 Servlet

以 Tomcat 中间件为例，Tomcat 的默认 Servlet 为 DefaultServlet，在 web.xml 配置文件中加入下列配置即可调用。

```
<servlet-mapping>
     <servlet-name>default</servlet-name>
     <url-pattern>*.jpg</url-pattern>
</servlet-mapping>
<servlet-mapping>
     <servlet-name>default</servlet-name>
     <url-pattern>*.js</url-pattern>
</servlet-mapping>
<servlet-mapping>
     <servlet-name>default</servlet-name>
     <url-pattern>*.css</url-pattern>
</servlet-mapping>
```

### 14.2.2 使用<mvc:resources>

在 Spring MVC 的配置文件内加入下列配置也可解决静态文件被拦截的问题。

```
<!-- 静态资源映射 -->
<mvc:resources location="/WEB-INF/view/" mapping="view/**"/>
```

(1) mapping：映射。

(2) location：本地资源路径，注意必须是 webapp 根目录下的路径。

(3) **：表示映射 resources/下所有的 URL，包括子路径(即接多个/)。

### 14.2.3　使用<mvc:default-servlet-handler/>

本节主要介绍<mvc:default-servlet-handler/>的使用方法。在 Spring MVC 的配置文件内加入下列配置。

```
<mvc:default-servlet-handler/>
```

使用<mvc:default-servlet-handler/>之后会把静态资源的处理转交给各个中间件自己默认的 Servlet 来处理。

## 14.3 项目编写

### 14.3.1　配置文件

引入静态文件，使用 Spring MVC 的 mvc:resources 标签处理。

```
<!--前缀后缀页面解析器-->
<bean id="viewResolver" class="org.springframework.web.servlet.view.InternalResourceViewResolver">
        <property name="suffix" value=".html"/>
        <property name="order" value="0"/>
        <property name="contentType" value="text/html;charset=UTF-8"></property>
    </bean>
    <!--静态资源映射-->
    <mvc:resources location="/WEB-INF/js/" mapping="js/**"/>
    <mvc:resources location="/WEB-INF/css/" mapping="css/**"/>
    <mvc:resources location="/" mapping="/favicon.ico"/>
```

### 14.3.2　引入静态文件

引入 JQuery、HTML 等静态文件，如图 14-3 所示。

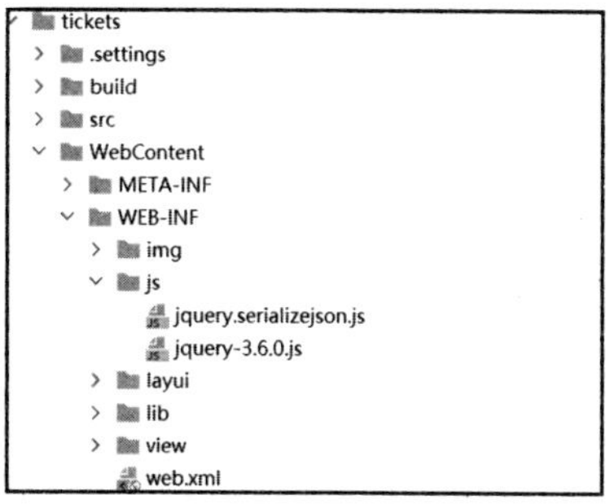

```
tickets
  > .settings
  > build
  > src
  v WebContent
    > META-INF
    v WEB-INF
      > img
      v js
          jquery.serializejson.js
          jquery-3.6.0.js
      > layui
      > lib
      > view
        web.xml
```

图 14-3

### 14.3.3 编写业务控制

MVC 模式的 View 的跳转由 Controller 层控制，登录和注册的代码具体如下。

```
/**
    * 返回登录页
    * @return 登录页
    * /
     @GetMapping("login")
      public String login() {
            return "login";
      }
  /**
    * 返回登录页
    * @return 登录页
    */
      @GetMapping("/user/login")
      public String userLogin() {
      return "login";
      }
/**
     * 返回注册页
     * @return 注册页
     */
    @GetMapping("regist")
    public String regist() {
          return "regist";
    }
```

### 14.3.4 登录页面

有了控制类，需要有具体的展示页面，登录页面的 HTML 部分核心源码如下。

```
<body>
<form id="form1">
      用户名：<input name="uname" type="text"/><br/>
      密码：<input name="pwd" type="password"/><br/>
      <button type="button" id="login">登录</button>
      <a href="regist">注册</a>
</form>
</body>
```

以上代码编写了一个 id 为“form1”的表单，同时有一个“登录”按钮和“注册”的跳转链接，如图 14-4 所示。

图 14-4

给页面引入 JavaScript 文件，编写 JavaScript 脚本，使用 AJAX 与后台交互。

```
<script type="text/javascript" src="js/jquery-3.6.0.js"></script>
<script>
 $(function(){
 //登录功能
 $("#login").click(){
        $.ajax({
                url: "test.html",                    //请求路径
                data: $("#form1").serialize(),
                type:"POST",                         //请求方式
                success: function(data){
                        alert(data);                 //后台给的返回结果
                }
        });
 }
 })
</script>
```

## 14.3.5　注册页面

注册页面的 HTML 与登录页面的类似部分的核心源码如下。

```
<form id="form1">
用户名：<input name="uname" type="text"/><br/>
密码：<input name="pwd" type="password"/><br/>
确认密码：<input name="pwd" type="password"/><br/>
性别：<input name="sex" type="radio" value="男"/>男<input name="sex" type="radio" value="女"/>男<br/>
电话：<input name="tel" type="text"/><br/>
<button type="button" id="regist">免费注册</button>
<a href="login">已有账号登录</a>
</form>
```

给页面引入 JavaScript 文件，编写 JavaScript 脚本，使用 AJAX 与后台交互。

```
<script type="text/javascript" src="js/jquery-3.6.0.js"></script>
<script>
$(document).ready(function(){
 //注册功能
```

```
        $("#login").click(function(){
                $.ajax({
                        url: "/tickets/user/regist",                //请求路径
                        data: $("#form1").serialize(),
                        type: "POST",                               //请求方式
                        contentType:"application/json",
                        success: function(data){
                                alert(data);                        //后台给的返回结果
                        }
                });
        });
        })
</script>
```

## 14.3.6 测试项目

访问登录页面，并且输入数据，单击“登录”按钮，如图 14-5 所示。

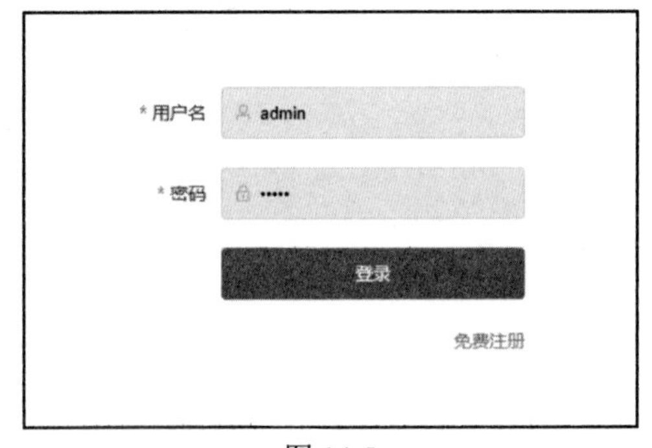


图 14-5

- 本单元开始有了前、后台的交互，只需要完成最基本的交互，JavaScript 的页面跳转大多是后台直接控制，业务逻辑耦合性非常强。
- 本单元展示了如何使用 AJAX 访问后台，继而后台响应前台的请求，给一个反馈信息或数据，然后前台根据后台反馈的信息或数据，进行业务逻辑的判断，做出对应的信息处理，如跳转、提示操作等。

# 页面美化

## 课程目标

❖ 掌握 UI 框架的使用

❖ 掌握拦截器的应用

❖ 掌握订单管理模块的设计方法

**简介**

前面几个单元，完成了项目的构建、业务的抽象化、基础静态文件的引入，并且使用 HTML 和 AJAX 与后台的 API 进行了简单的交互，完成了最基础的由 VIEW 层传递数据至 DataBase 的操作。本单元将在前几单元的基础上，完成 UI 框架的使用、拦截器的应用及订单管理模块的设计。

# 15.1 UI 框架的使用

当下很多 UI 框架，例如 BootStrap、EasyUI、Amaze UI、LayUI 等被广泛使用，“百里半门票销售系统”采用了其中的一种，使用 LayUI 做为后台管理的展示框架。

## 15.1.1 业务功能图

对登录页面做美化，效果如图 15-1 所示。

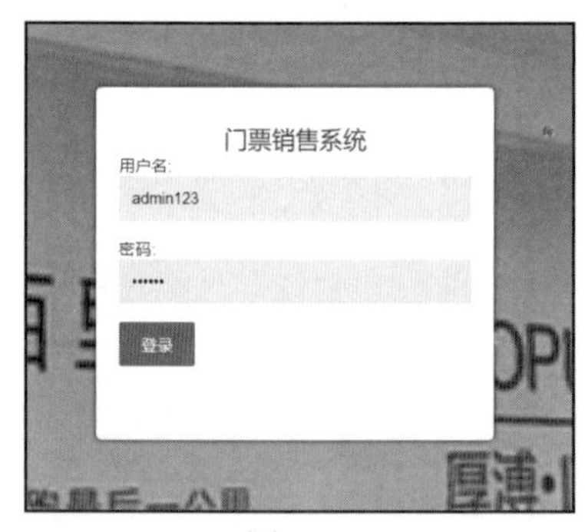


图 15-1

订单展示列表如图 15-2 所示。

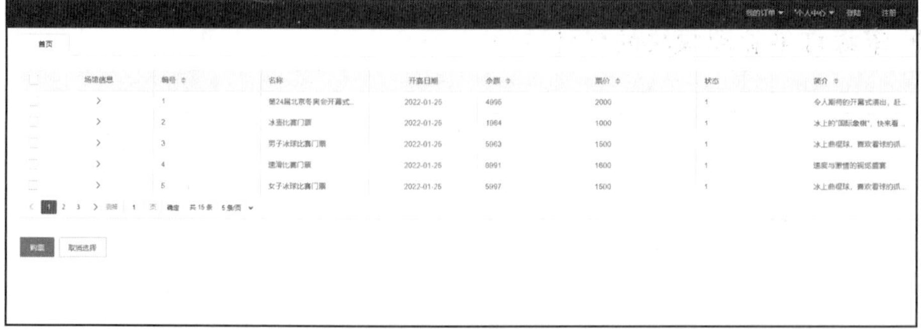

图 15-2

## 15.1.2　引入 UI 框架

首先在页面内引入 CSS 文件和 JS 文件。

在页面内引入 LayUI 的 CSS 文件和 JS 文件。

```
<link rel="stylesheet" href="/tickets/layui/css/layui.css">
<script src="/tickets/layui/layui.js"></script>
```

然后在 JavaScript 内配置 LayUI 的使用，就可以使用 LayUI 框架了。

```
//JavaScript 代码区域
//如果想要使用 JQuery，需要使用 $ = layui.jquery 调用 LayUI 内封装的 Jquery。
layui.use(['jquery', 'laydate', 'util', 'layer', 'table', 'form'], function(){
    var table = layui.table , $ = layui.jquery , form = layui.form, layer = layui.layer;
});
```

## 15.1.3　UI 框架的基本使用

以修改登录页为例，首先引入 LayUI 的 JS 文件和 CSS 文件。

```
<script src="/tickets/layui/layui.js"></script>
<link rel="stylesheet" href="/tickets/layui/css/layui.css">
```

继续使用 AJAX，这里依然可以按照 JQuery 的方式使用。

写法如下：需要单独列出 JQuery，从 LayUI 中获取。

```
layui.use(['jquery'], function () {
        var $ = layui.jquery;
        //登录功能
         $("#login").click(function(){
            $.ajax({
                url: "/tickets/user/login", //请求路径
                data: JSON.stringify($("#form1").serializeJSON()),
                type: "POST",//请求方式
                contentType:"application/json",
                dataType: "json",
                success: function(data){
                    if(data.code != 200){
                        alert(data.message);
                    }
                    if(data.code == 200){
                        alert(data.message);
                        location.href="/tickets/index";
                    }
                }
            });
```

```
            });
    })
```

# 15.2 门票列表

## 15.2.1 对应 API

从门票列表页可以看出，有一个列表形式展示的页面(如图 15-3 所示)，下边有页码、页数、列表三部分。

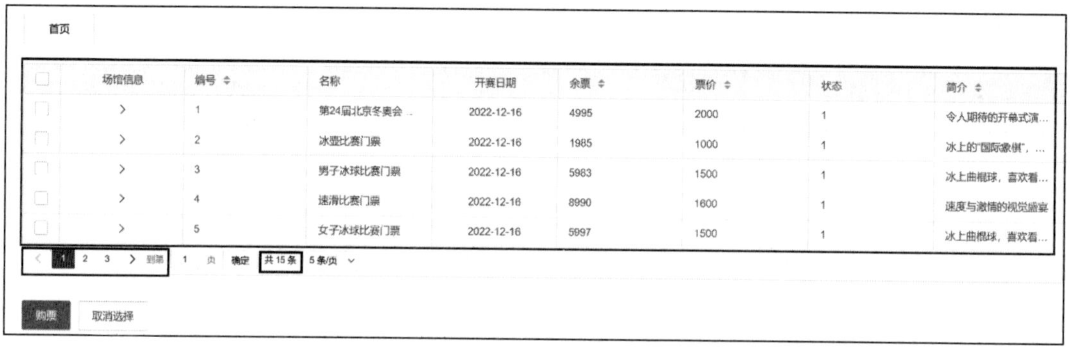

首页

| | 场馆信息 | 编号 | 名称 | 开赛日期 | 余票 | 票价 | 状态 | 简介 |
|---|---|---|---|---|---|---|---|---|
| | > | 1 | 第24届北京冬奥会 ... | 2022-12-16 | 4995 | 2000 | 1 | 令人期待的开幕式演... |
| | > | 2 | 冰壶比赛门票 | 2022-12-16 | 1985 | 1000 | 1 | 冰上的"国际象棋"，... |
| | > | 3 | 男子冰球比赛门票 | 2022-12-16 | 5983 | 1500 | 1 | 冰上曲棍球，喜欢看... |
| | > | 4 | 速滑比赛门票 | 2022-12-16 | 8990 | 1600 | 1 | 速度与激情的视觉盛宴 |
| | > | 5 | 女子冰球比赛门票 | 2022-12-16 | 5997 | 1500 | 1 | 冰上曲棍球，喜欢看... |

< 1 2 3 > 到第 1 页 确定 共 15 条 5 条/页

购票 取消选择

图 15-3

列表部分显示了具体的内容，包括门票的名称、门票的日期、门票的编号、剩余数量等。而列表的下边显示了具体总共有多少页，总共有多少种门票，也就是门票的总条目数。

图 15-3 显示的数量仅有 15 条，还比较少，展示能全部展示在同一个页面，但是如果数量继续增加，达到了 150 条、1500 条、15000 条等，还能在同一个页面展示么？显然是不可以的。那么列表就必须按照上图的形式来展示。这种展示形式利用了最常见的展示方式——数据的分页处理。

## 15.2.2 列表分页

日常软件开发过程中常常把大量的数据以固定顺序、固定大小的形式分割成若干份，然后通过下标的形式展示于前台页面。

例如，把 1000 条数据以 10 条每页的形式分割成 100 个页面，当下标为 0 的时候把第 1 条数据至第 10 条数据展示于前台，当下标为 1 的时候把第 11 条数据至第 20 条数据展示于前台页面，依次类推，直至最后一个页面。

如果想要完成这样的效果，一般都是由前后台共同参与，然后一起协同最后完成分页的展示。前台页面基本都是图 15-3 的三个部分，而后台 API 需要接收到的参数必须有具体的下标和每页需要展示的数量。

后台分页的 API 如下：

```
http://地址/项目名字/show/页数/每页的数量
http://127.0.0.1/tickets/ticket/show/1/5
```

接口控制层的源码如下：

```
@GetMapping("/show/{page}/{size}")
public Result findAll(@PathVariable("page") int page, @PathVariable("size") int size) {
PageInfo pageInfo = ticketService.findAll(page, size);
return Result.success(pageInfo);
}
```

上边 API 参数中的 page 表示需要展示的页码、size 表示每页的数量，通过后台的计算得出应该在前台展示从(page-1) * size 开始至 page * size 结束的数据。

MySql 的 SQL 语句如下：

```
SELECT  *  FROM  TABLE_NAME  LIMIT (page-1) * size , size
```

返回的数据内容如下：

```
{
  "code": 200,
  "message": "成功",
  "success": **true**,
  "data": {
    "total": 15,
    "list": [...],
    "pageNum": 1,
    "pageSize": 5,
    "size": 5,
    "startRow": 1,
    "endRow": 5,
    "pages": 3,
    "prePage": 0,
    "nextPage": 2,
    "isFirstPage": **true**,
    "isLastPage": **false**,
    "hasPreviousPage": **false**,
    "hasNextPage": **true**,
    "navigatePages": 8,
    "navigatepageNums": [
      1,
      2,
```

```
            3
        ],
        "navigateFirstPage": 1,
        "navigateLastPage": 3
    }
}
```

可以看出以上的数据格式是标准的 jsonObject 形式的数据库格式，其中的一个叫作 list 的 key 是 jsonArray 的数据格式，list 内存放的是具体的数据内容。“百里半门票销售系统”采用了 PageHelper 插件。

### 15.2.3 PageHelper 插件

在 Spring 的配置文件或者 MyBatis 的配置文件中加入配置。

在 Spring 中加入配置，代码如下：

```
<!--MyBatis 配置-->
  <bean id="sqlSessionFactory" class="org.mybatis.spring.SqlSessionFactoryBean">
          <property name="dataSource" ref="dataSource"/>
          <property name="configLocation" value="classpath:config/mybatis-config.xml"/>
          <!--MyBatis 插件-->
          <property name="plugins">
               <set>
                    <!--配置 pageHelper 分页插件-->
                    <bean class="com.github.pagehelper.PageInterceptor">
                         <property name="properties">
                              <props>
                                   <!--方言：-->
                                   <prop key="helperDialect">mysql</prop>
                                   <prop key="reansonable">true</prop>
                              </props>
                         </property>
                    </bean>
               </set>
          </property>
     </bean>
```

在 MyBatis 中增加配置，代码如下：

```
<!-- MyBatis 拦截器(插件) -->
     <plugins>
      <plugin interceptor="com.github.pagehelper.PageInterceptor">
           <property name="helperDialect" value="mysql"/>
           <property name="reansonable" value="true"/>
      </plugin>
     </plugins>
```

对 PageHelper 参数解释如下。

- pageNum：表示当前页码。
- pageSize：表示每页个数。
- size：表示当前页个数。
- startRow：表示由第几条开始。
- endRow：表示到第几条结束。
- total：表示总条数。
- pages：表示总页数。
- list：表示查出的数据集合。
- prePage：表示上一页。
- nextPage：表示下一页。
- isFirstPage：表示是否为首页。
- isLastPage：表示是否为尾页。
- hasPreviousPage：表示是否有上一页。
- hasNextPage：表示是否有下一页。
- navigatePages：表示每页显示的页码个数。
- navigateFirstPage：表示首页。
- navigateLastPage：表示尾页。
- navigatepageNums：表示页码数。

### 15.2.4　门票列表思路

(1) 使用 AJAX 访问后台 API。

(2) 后台 API 返回 JSON 串。

(3) 拿到 JSON 串解析 JSON 串。

(4) 使用 JavaScript 把 JSON 串以想要的样式展示于页面之上。

不使用 UI 框架写法。

#### 1. 访问和返回

前台使用 AJAX 访问后台的 API，AJAX 的 success 是后台返回的数据。代码如下：

```
 function ready(){
$(".active").click(function(){
   $.ajax({
         url: "/tickets/ticket/show/"+$(this).text()+"/5",
         type: "GET",//请求方式
```

```
                contentType:"application/json",
                dataType: "json",
                success: function(data){
                    //这个 data 是后台返回的数据
                     let tr="";
                     $.each(data.data.list,function(i,item){
                         tr =tr + trs(item);
                    });
                    $("#tables").html(tr);
                    $("#tablefoot").html(ps(data.data));
                    $(".active").hover(function(){
                              $(this).css('cursor','pointer')
                     },function(){
                                        $(this).css('cursor','')
                     });
                     ready();
                }
            });
        });
}
```

解析后台返回的 JSON 数据，使用 JavaScript 处理。代码如下：

```
$.each(data.data.list,function(i,item){
    tr =tr + trs(item);
});
$("#tables").html(tr);
$("#tablefoot").html(ps(data.data));
$(".active").hover(function(){
    $(this).css('cursor','pointer')
},function(){
    $(this).css('cursor','')
});
```

### 2. 解析和展示

利用 JQuery 的 html()方法直接改变 HTML 代码重新渲染。

表格内容如图 15-4 所示。

首页

|  | 场馆信息 | 编号 | 名称 | 开赛日期 | 余票 | 票价 | 状态 | 简介 |
|---|---|---|---|---|---|---|---|---|
|  | > | 1 | 第24届北京冬奥会… | 2022-12-16 | 4995 | 2000 | 1 | 令人期待的开幕式演… |
|  | > | 2 | 冰壶比赛门票 | 2022-12-16 | 1985 | 1000 | 1 | 冰上的"国际象棋"，… |
|  | > | 3 | 男子冰球比赛门票 | 2022-12-16 | 5983 | 1500 | 1 | 冰上曲棍球，喜欢看… |
|  | > | 4 | 速滑比赛门票 | 2022-12-16 | 8990 | 1600 | 1 | 速度与激情的视觉盛宴 |
|  | > | 5 | 女子冰球比赛门票 | 2022-12-16 | 5997 | 1500 | 1 | 冰上曲棍球，喜欢看… |

< 1 2 3 > 到第 1 页 确定 共 15 条 5 条/页

购票 取消选择

图 15-4

代码如下：

```
function trs(params){
    var tr=""
    +" <tr class=\"el-table__row\">"
    +" <td rowspan=\"1\" colspan=\"1\" class=\"el-table_1_column_1 el-table-column--selection\">"
    +" <div class=\"cell\"><label class=\"el-checkbox\"><span class=\"el-checkbox__input\">
       <span class=\"el-checkbox__inner\"></span><input"
    +" type=\"checkbox\" aria-hidden=\"false\" class=\"el-checkbox__original\" value=\"
       "+params.tid+"\"></span>"
    +" </label></div>"
    +" </td>"
    +" <td rowspan=\"1\" colspan=\"1\" class=\"el-table_1_column_2
       undefined el-table__expand-column\">"
    +" <div class=\"cell\">"
    +" <div class=\"el-table__expand-icon\"><i class=\"el-icon el-icon-arrow-right\"></i></div>"
    +" </div>"
    +" </td>"
    +" <td rowspan=\"1\" colspan=\"1\">"
    +" <div class=\"cell\">"+params.tid+"</div>"
    +" </td>"
    +" <td rowspan=\"1\" colspan=\"1\">"
    +" <div class=\"cell\">"+params.tname+"</div>"
    +" </td>"
    +" <td rowspan=\"1\" colspan=\"1\">"
    +" <div class=\"cell\"><i class=\"el-icon-time\"></i> <span style=\"margin-left: 10px;\">
       1643904000000</span></div>"
    +" </td>"
    +" <td rowspan=\"1\" colspan=\"1\">"
    +" <div class=\"cell\">"+params.ticketRemainNum+"</div>"
    +" </td>"
    +" <td rowspan=\"1\" colspan=\"1\">"
    +" <div class=\"cell\">"+params.price+"</div>"
    +" </td>"
    +" <td rowspan=\"1\" colspan=\"1\">"
    +" <div class=\"cell\">"+params.state+"</div>"
    +" </td>"
    +" <td rowspan=\"1\" colspan=\"1\">"
    +" <div class=\"cell\">"+params.description+"</div>"
    +" </td>"
    +" </tr>";
    return tr;
}
```

页码和页数如图 15-5 所示。

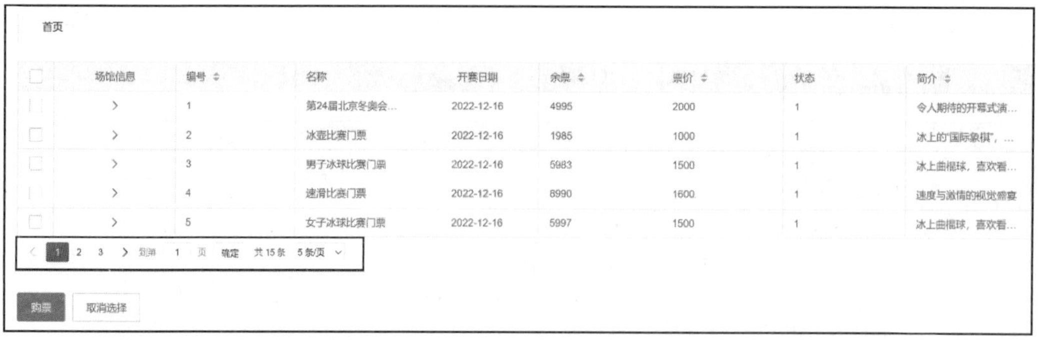

首页

| | 场馆信息 | 编号 | 名称 | 开赛日期 | 余票 | 票价 | 状态 | 简介 |
|---|---|---|---|---|---|---|---|---|
| | > | 1 | 第24届北京冬奥会... | 2022-12-16 | 4995 | 2000 | 1 | 令人期待的开幕式演... |
| | > | 2 | 冰壶比赛门票 | 2022-12-16 | 1985 | 1000 | 1 | 冰上的"国际象棋"，... |
| | > | 3 | 男子冰球比赛门票 | 2022-12-16 | 5983 | 1500 | 1 | 冰上曲棍球，喜欢看... |
| | > | 4 | 速滑比赛门票 | 2022-12-16 | 8990 | 1600 | 1 | 速度与激情的视觉盛宴 |
| | > | 5 | 女子冰球比赛门票 | 2022-12-16 | 5997 | 1500 | 1 | 冰上曲棍球，喜欢看... |

< 1 2 3 > 到第 1 页 确定 共 15 条 5 条/页

购票 取消选择

图 15-5

代码如下：

```
function pages(pageNums,pagess){
    let pageNum = parseInt(pageNums);
    let pages =   parseInt(pagess);
    let pn="";
    if(pageNum<3&&pages<=2){                    //判断页面是否有前两页
        pn=pn+"<li class=\"number active\">"+(pageNum)+"</li>";
        pn=pn+"<li class=\"number active\">"+(pageNum+1)+"</li>";
        return pn;
    }
    if((pageNum-2)>0){                          //判断页面是否有前两页
        pn=pn+"<li class=\"number active\">"+(pageNum-2)+"</li>";
    }
    if((pageNum-1)>0){                          //判断页面是否有前一页
        pn=pn+"<li class=\"number active\">"+(pageNum-1)+"</li>";
    }
    pn=pn+"<li class=\"number active\"><font color=\"black\">"+(pageNum)+"</font></li>";
    if((pageNum+1)<=pages){                     //判断页面是否有后一页
        pn=pn+"<li class=\"number active\">"+(pageNum+1)+"</li>";
    }
    if((pageNum+2)<=pages){                     //判断页面是否有后两页
        pn=pn+"<li class=\"number active\">"+(pageNum+2)+"</li>";
    }
    return pn;
}
```

## 15.2.5 使用 UI 框架完成设计

流程思路与上边部分一致，首先使用前台访问后台的接口，然后展示于前台的页面。切换 LayUI 后代码如下：

```
layui.use(['jquery', 'layer', 'table', 'form'], function(){
  var table = layui.table,                    //获取 table 的使用
```

```
        $ = layui.jquery,                    //获取 JQuery 的使用
        form = layui.form,                   //获取 form 的使用
        layer = layui.layer;                 //获取 layer 弹窗的使用
    table.render({
      elem: '#test'
      ,id:'#tabletest'
      ,url:'/tickets/ticket/show'
      ,cellMinWidth: 80                      //全局定义常规单元格的最小宽度，LayUI 2.2.1 新增
      ,limit:5
      ,limits:[5,10,20,50]
      ,cols: [[
        {type:'checkbox'}
        ,{field:'tid', title: '场馆信息',
        templet:function(d){return "<i class=\"layui-icon\" lay-event=\"addRowTable\">&#xe602;</i>";},
        align:'center'}
        ,{field:'tid', title: '编号', sort: true}
        ,{field:'tname', title: '名称'}
        ,{field:'begintime', title: '开赛日期', templet : "<div>{{layui.util.toDateString(d.rTime,
          'yyyy-MM-dd')}}</div>", align:'center'}
        ,{field:'ticketRemainNum', title: '余票', sort: true}
        ,{field:'price', title: '票价', sort: true}
        ,{field:'state', title: '状态'}
        ,{field:'description',minWidth:150,   title: '简介', sort: true}
      ]]
      ,page: true
      ,done: function(res){
            tableDatas = res.data;           //返回结果封装的数据集
      }
    });
});
```

## 单元小结

- 经过三个单元，对项目有了一定的认识，对前后端的交互有了基础的理解，后端返回的 JSON 数据在前台落地，后台响应前台的请求之后给出 JSON 数据，前台需要根据反馈的数据进行展示等，需要使用大量的 JavaScript 代码。
- 本单元开始深入理解后台仅仅是一个响应端口，具体展示 JavaScript 要依靠 View 而不是 Controller，页面的变化需要依靠 JavaScript 而不是 Server 端。Java 并不能对前端数据的处理给出帮助，不能像 JSP 一样使用各种标签直接循环输出结果，而 JavaScript 的使用又特别麻烦，所以市场中有很多 Html 的框架。

- 一般做项目仅仅要求会使用即可，只有一些特别定义的样式需要进行一定的自定义，其余内容几乎都是复制、粘贴，把前端框架引入项目，跟之前的接口对应起来，完成项目速度会非常快。

# 复用与结尾

## 课程目标

- ❖ 掌握功能的复用
- ❖ 掌握拦截器的实际应用
- ❖ 掌握购票模块的设计方法
- ❖ 掌握个人中心模块的设计方法

**简介**

上一单元引入了比较简单的前端框架 LayUI，介绍了分页操作，使用了分页的插件，在本单元将继续完成对应的操作，复用上一单元所学习的内容，完成对应的功能。除此之外本单元还会将 Spring MVC 的拦截器应用于实际的项目之中。

# 16.1 功能的复用

还原任何项目至最基础的操作都是增、删、查、改这 4 种操作，也就是 CURD。页面大多数都是对数据进行展示，做列表的处理。大部分功能基本都是相同的操作，只是字段名展示的名称不同，但是后台的基础和业务都很类似，所以常常会有把其他功能复用至当下需要完成的业务。这里以“我的订单”为实例来介绍，同时在原有基础上增加一些变化条件。

## 16.1.1 业务功能图

单击“我的订单”中的“已支付”“待支付”“已取消”3 个选项(如图 16-1 所示)，在前台看起来是 3 种毫无关联的订单，但是在后台数据库中，对订单而言是 3 种不同的状态，这个状态可以使用一个字段来表示。

例如，水在 0 度以下是冰、0 度以上 100 度以下是液体、100 度以上是水蒸气。本质还都是 $H_2O$，仅是视觉看起来不同。对于后台来说都是查询后台的同一个数据库，仅仅是条件不同展示给前台返回的内容就有不同。

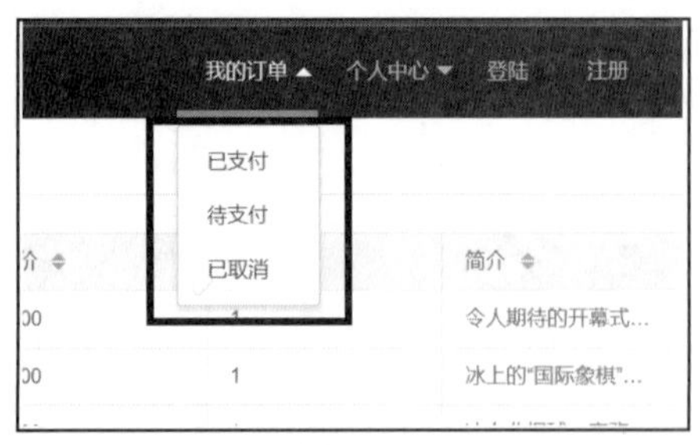


图 16-1

## 16.1.2 View 层源码

View 层源码展示了 3 个不同的页面跳转。

```
<li class="layui-nav-item">
        <a href="javascript:;">
          我的订单
        </a>
        <dl class="layui-nav-child">
          <dd><a href="/tickets/order/pay-order">已支付</a></dd>
          <dd><a href="/tickets/order/unpaid-order">待支付</a></dd>
          <dd><a href="/tickets/order/cancle-pay">已取消</a></dd>
        </dl>
</li>
```

“已支付”的订单页面：首先访问后台 API，然后后台 API 返回数据，页面进行展示。代码如下：

```
table.render({
    elem: '#test'
    ,url:'/tickets/orderBase/0'
    ,cellMinWidth: 80                          //全局定义常规单元格的最小宽度，LayUI 2.2.1 新增
    ,limit:5
    ,limits:[5,10,20,50]
    ,cols: [[
      {type:'checkbox'}
      ,{field:'id', title: '门票详情',templet:function(d){return "<i class=\"layui-icon\"
      lay-event=\"addRowTable\">&#xe602;</i>";}, align:'center'}
      ,{field:'obid', title: '订单编号', sort: true}
      ,{field:'ordertime', title: '下单时间'}
      ,{field:'paytime', title: '付款时间', templet : "<div>{{layui.util.toDateString(d.rTime,
      'yyyy-MM-dd')}}</div>", align:'center'}
      ,{field:'totalNum', title: '门票数量', sort: true}
      ,{field:'sumPrice', title: '实付金额', sort: true}
      ,{fixed: 'right', title:'操作', toolbar: '#barDemo', width:160}
    ]]
    ,page: true
    ,done: function(res){
          tableDatas = res.data;                //返回结果封装的数据集
    }
  });;
```

“待支付”的订单页面：首先访问后台 API，然后后台 API 返回数据，页面进行展示。

```
table.render({
    elem: '#test'
    ,url:'/tickets/orderBase/0'
    ,cellMinWidth: 80                          //全局定义常规单元格的最小宽度，LayUI 2.2.1 新增
    ,limit:5
    ,limits:[5,10,20,50]
    ,cols: [[
      {type:'checkbox'}
```

```
            ,{field:'id', title: '门票详情',templet:function(d){return "<i class=\"layui-icon\"
            lay-event=\"addRowTable\">&#xe602;</i>";}, align:'center'}
            ,{field:'obid',   title: '订单编号', sort: true}
            ,{field:'ordertime',   title: '下单时间'}
            ,{field:'paytime', title: '付款时间', templet : "<div>{{layui.util.toDateString(d.rTime,
            'yyyy-MM-dd')}}</div>", align:'center'}
            ,{field:'totalNum',   title: '门票数量', sort: true}
            ,{field:'sumPrice',   title: '实付金额', sort: true}
            ,{fixed: 'right', title:'操作', toolbar: '#barDemo', width:150}
        ]]
        ,page: true
        ,done: function(res){
              tableDatas = res.data;                    //返回结果封装的数据集
        }
    });
```

“已取消”的订单页面：首先访问后台 API，然后后台 API 返回数据，页面进行展示。

```
    table.render({
        elem: '#test'
        ,id:'#tabletest'
        ,url:'/tickets/orderBase/-1'
        ,cellMinWidth: 80                        //全局定义常规单元格的最小宽度，LayUI 2.2.1 新增
        ,limit:5
        ,limits:[5,10,20,50]
        ,cols: [[
            {type:'checkbox'}
            ,{field:'id', title: '门票详情',templet:function(d){return "<i class=\"layui-icon\"
            lay-event=\"addRowTable\">&#xe602;</i>";}, align:'center'}
            ,{field:'obid', title: '订单编号', sort: true}
            ,{field:'ordertime', title: '下单时间'}
            ,{field:'paytime', title: '付款时间', templet : "<div>{{layui.util.toDateString(d.rTime,
            'yyyy-MM-dd')}}</div>", align:'center'}
            ,{field:'totalNum', title: '门票数量', sort: true}
            ,{field:'sumPrice', title: '实付金额', sort: true}
        ]]
        ,page: true
        ,done: function(res){
              tableDatas = res.data;                    //返回结果封装的数据集
        }
      });;
```

以上 3 个页面几乎一致，同时与前一个章节的门票展示也几乎一致，唯一不同的是这里的 3 个不同的业务同时对应同一个 API，通过向后台传递不同的参数代表不同的状态，展示不同的数据。

### 16.1.3　API 的描述

“我的订单”对应的 API 如下：

```
/tickets/orderBase/{state}
```

使用/{state}接收不同的数据。与前边的门票列表订单类似，但是这里关联了下单的信息和用户关联。继续使用@PathVariable 注解通过不同的状态控制订单信息。

```
/**
 * 查询指定状态下用户的全部订单
 *
 * @param session session
 * @param state     0：待支付；1：已支付；-1：已取消
 * @param page      页码
 * @param size      每页条数
 * @return Result
 */
@GetMapping("/{state}")
public Result queryOrdersByUidAndState(HttpSession session, @PathVariable("state") Integer state
        ,@RequestParam("page") Integer page
        , @RequestParam("limit") Integer limit) {
    // 获取当前用户
    User user = (User) session.getAttribute("user");
    PageInfo<OrderBase> pageInfo = orderBaseService.queryOrdersByUidAndState(user.getUid(),
            state, page, limit);
    return Result.success(pageInfo.getList(),pageInfo.getTotal());
}
```

## 16.2 拦截器的实际应用

### 16.2.1　业务需求

假如在未登录的情况下就访问个人信息、已支付、未支付订单，那么究竟是谁访问的已经支付的订单和未支付的订单？

正常来说未登录用户不被允许直接访问订单、修改用户信息等已登录用户才能操作的内容。单击之后返回登录界面(如图 16-2 所示)，这里可以使用拦截器来完成这一功能。

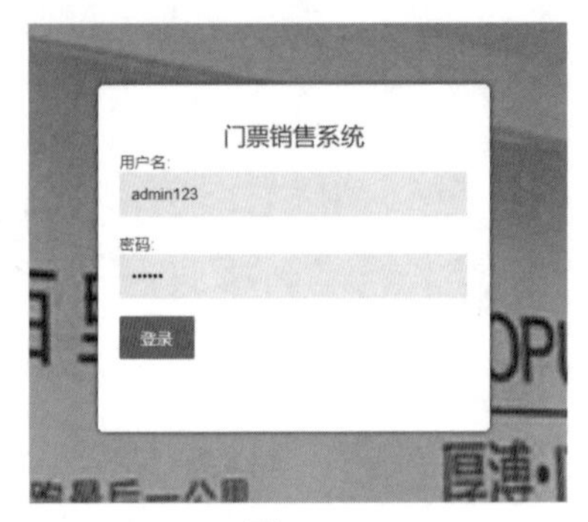


图 16-2

## 16.2.2 业务实现

给 Spring MVC 加上控制器，代码如下。

```
<!--配置拦截器-->
<mvc:interceptors>
    <mvc:interceptor>
        <!--需要拦截的 url 路径 :/** -->
        <mvc:mapping path="/**"/>
        <!-- mvc:exclude-mapping 是另外一种拦截，它可以在后来的测试中对某个页面不进行
         拦截，这样就不用在 LoginInterceptor 的 preHandler 方法里面获取不拦截的请求 uri
         地址了(优选) -->
        <mvc:exclude-mapping path="/"/>
        <mvc:exclude-mapping path="/login"/>
        <mvc:exclude-mapping path="/favicon.ico"/>
        <mvc:exclude-mapping path="/user/login"/>
        <mvc:exclude-mapping path="/user/logout"/>
        <mvc:exclude-mapping path="/user/regist"/>
        <mvc:exclude-mapping path="/user/isLogin"/>
        <mvc:exclude-mapping path="/regist"/>
        <mvc:exclude-mapping path="/index"/>
        <mvc:exclude-mapping path="/ticket/show/**"/>
        <mvc:exclude-mapping path="/js/**"/>
        <mvc:exclude-mapping path="/css/**"/>
        <bean id="loginInterceptor" class="com.bailiban.interceptor.LoginInterceptor"/>
    </mvc:interceptor>
</mvc:interceptors>
```

拦截器需要实现 HandlerInterceptor 接口，代码逻辑如下：

```
public class LoginInterceptor implements HandlerInterceptor {
    @Override
    public boolean preHandle(HttpServletRequest request, HttpServletResponse response, Object handler)
    throws Exception {
        HttpSession session = request.getSession();
// 用户已登录，放行
```

```
        if (session.getAttribute("user") != null) {
            return true;
        }
        String requestUri = request.getRequestURI();
// 用户新注册，允许访问身份校验页以及发送身份绑定请求
        if (session.getAttribute("registOK") != null) {
            if (requestUri.contains("checkCard") || requestUri.contains("bindCard")) {
                return true;
            }
        }
// 用户没有登录跳转到登录页面
        System.out.println("未登录，跳转到登录页");
        response.sendRedirect("/ticket_sales_system/login");
        return false;
    }

}
```

## 16.3 购票

### 16.3.1 业务功能图

在上一单元已经完成了门票查询的操作，这里对这一部分业务稍加完善，增加购票操作。首先勾选需要购买的门票，然后单击“购买”，如图 16-3 所示。

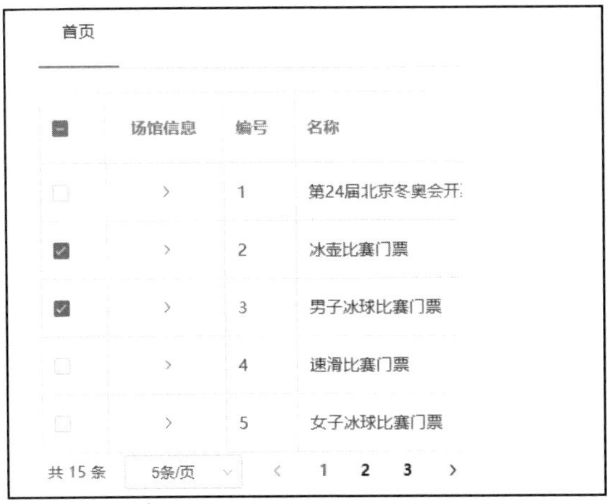

图 16-3

单击“购买”后弹出“下单”窗口，如图 16-4 所示。

图 16-4

## 16.3.2 View 层源码

弹出“下单”窗口的功能代码如下。

```
$("#buyBtn").click(function(){
            tableData = new Array();
            $(".laytable-cell-checkbox").each(function(i,item){
        //由于 google 5.8 不支持使用 getCheckstatus，所以换一种写法
        //item 对象中有两个元素，一个是 checkbox，另一个是包含选中与否的样式的 div
        if($(item).find(".layui-form-checked").length>0){
            tableData.push(tableDatas[i]);
        }
    });
    layer.open({
        type: 2 ,
        title: '下单',
        area: ['800px', '400px'],                          //宽高
        shade: 0.4,                                        //遮罩透明度
        content: "/tickets/tickets",                       //支持获取 DOM 元素
        scrollbar: false ,                                 //屏蔽浏览器滚动条
        yes: function(index){//layer.msg('yes');           //单击“确定”回调
            layer.close(index);
            showToast();
        },
        btn2: function(){//layer.alert('aaa',{title:'msg title'});    //单击“取消”回调
        layer.msg('bbb');//layer.closeAll();
        }
    });
});
```

## 16.3.3 API 的描述

把前台提交的数据交给后台的 API 处理，代码如下。

```
@PostMapping("/add")
public Result order(HttpSession session, @RequestBody List<OrderParam> orders) {
            System.out.println(orders);
```

```
        double sumPrices = 0;
        int nums = 0;
        List<OrderDetail> orderDetails = new ArrayList<>();
        for (OrderParam order : orders) {
            Ticket ticket = ticketService.queryById(order.getTid());
            nums += order.getNum();
            System.out.println("门票信息：" + ticket);
            System.out.println("订单参数信息：" + order);
            sumPrices += order.getNum() * ticket.getPrice();
        // 插入订单详情，稍后补充 oid
            System.out.println("总价格：" + sumPrices);
            OrderDetail detail = new OrderDetail();
            detail.setGoodsName(ticket.getTname());
            detail.setGoodsNum(order.getNum());
            detail.setUnitPrice(ticket.getPrice());
            detail.setGoodsPrice(ticket.getPrice() * order.getNum());
            detail.setTid(order.getTid());
            orderDetails.add(detail);
        // 更新门票数量
            ticket.setTicketRemainNum(ticket.getTicketRemainNum() - order.getNum());
            ticketService.update(ticket);
        }
        OrderBase orderBase = new OrderBase();
        orderBase.setObid(UUID.randomUUID().toString().replaceAll("-", ""));
        orderBase.setOrdertime(LocalDateTime.now());
        orderBase.setState(0);
        orderBase.setSumPrice(sumPrices);
        orderBase.setTotalNum(nums);
        User user = (User) session.getAttribute("user");
        orderBase.setUid(user.getUid());
        // 插入订单基础信息
        orderBase = orderBaseService.insert(orderBase);
        // 完善插入订单详情，补充 id
        if (orderBase != null) {
            for (OrderDetail orderDetail : orderDetails) {
        orderDetail.setOid(orderBase.getId());
            }
        // 插入订单详情
            int i = orderDetailService.insertOrderDetails(orderDetails);
            if (i > 0) {
        return Result.success(orderBase.getId());
            }
        }
        return Result.failure(ResultCode.UNKNOWN_REASON);
}
```

# 16.4 个人中心

## 16.4.1 业务功能

个人中心有 3 个功能，分别是身份验证、个人信息、退出系统，如图 16-5 所示。

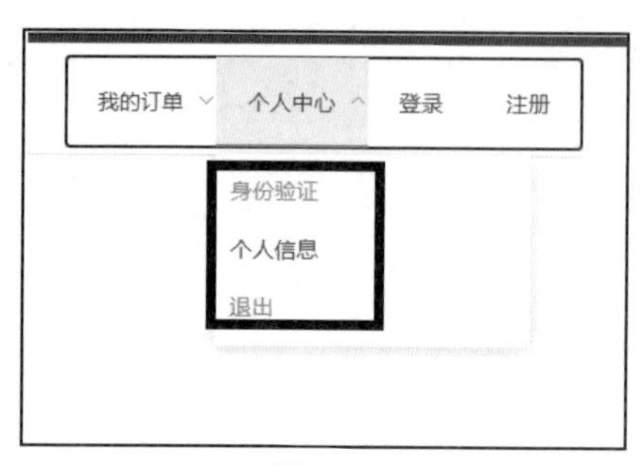


图 16-5

## 16.4.2 身份验证

“身份验证”是对注册信息的补充和完善，对于后台而言就是更新信息，如图 16-6 所示。

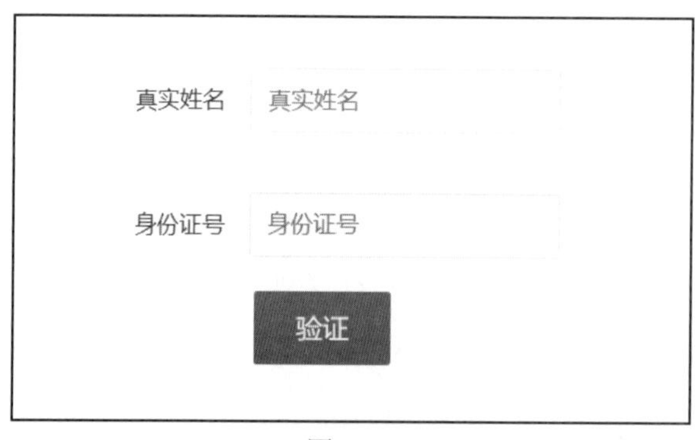


图 16-6

向后台数据提交信息的代码如下。

```
/**
 * 用户校验、更新证件，也就是完善用户的身份信息
 */
@PostMapping("bindCard")
    public Result bindCard(HttpSession session, @RequestBody(required = false) User user) {
        // 非空校验，当前 user 中只有真实姓名和身份证号
        if (user != null) {
        // 更新证件信息，可以在注册之后立刻校验绑定身份信息，也可以选择登录后再绑定
```

```
            User registOK = (User) session.getAttribute("registOK");
            User loginUser = (User) session.getAttribute("user");
            if (registOK != null) {
                // 如果是注册后绑定，此时处于未登录状态，从注册信息中获取 uid
                user.setUid(registOK.getUid());
            } else {
                // 否则从登录信息中获取 id
                user.setUid(loginUser.getUid());
            }
            // 更新信息，也就是绑定身份信息
            user = userService.update(user);
            // 绑定后重新更新 session 中的 user
            session.setAttribute("user", user);
            return Result.success();
        }
        return Result.failure(ResultCode.PARAM_IS_BLANK);
    }
```

## 16.4.3　个人信息

“个人信息”仅仅是对数据的查看，例如个人信息包含了哪些内容。此操作需要访问后台 API。

```
/user/myinfo
```

返回的 JSON 数据如下。

```
{"code":0,
"message":"成功",
"success":true,
"data":{
    "uid":6,
    "uname":"admin",
    "pwd":"admin",
    "realname":"张三",
    "idcard":"4101123412121234",
    "sex":"男",
    "tel":"13388888800",
    "registrationTime":"2020-09-16 16:07:05"
},
"count":null
}
```

使用 AJAX 的 each()函数，通过与标签的 name 对应，把数据一个个填充到元素内部。

```
$.ajax({
    url: "/tickets/user/myinfo",          //请求路径
```

```
        type: "GET",                        //请求方式
        success: function(data){
            $.each(data.data,function(i,item){
                $("input[name='"+i+"']").val(item);
            });
        }
});
```

### 16.4.4 退出系统

退出也被称为注销，在 Servlet 中的 HttpSession 有一个方法是:

```
invalidate();
```

使用这个方法可以销毁当前的 session。

```
@GetMapping("/exit")
public String exit(HttpSession session) {
session.invalidate();//注销
return "login";
}
```

## 单元小结

- CURD 复用至项目内的其他功能就是对业务操作的补充，常见的页面展示几乎都是页面列表的展示，所以可以把类似的功能直接修改复用至新的功能。
- 拦截器在项目内大多都是对业务功能的筛选，在本单元对用户是否已经登录做了最终的处理。
- 本单元最后完成了对业务的最后补充操作，完成了项目的业务功能、身份验证、个人信息、退出系统这几项基本业务。